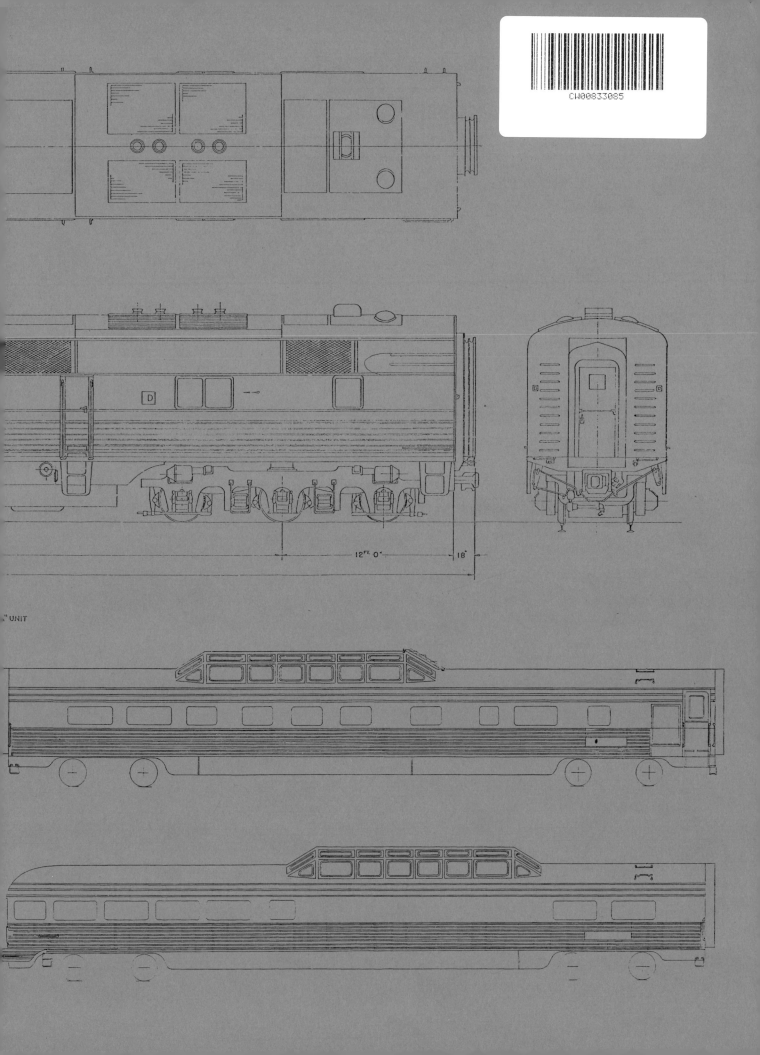

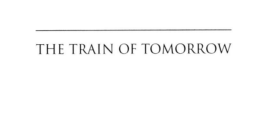
THE TRAIN OF TOMORROW

Railroads Past and Present
George M. Smerk, editor

THE TRAIN OF TOMORROW

RIC MORGAN

Indiana University Press
Bloomington and Indianapolis

This book is a publication of

Indiana University Press
601 North Morton Street
Bloomington, IN 47404-3797 USA

http://iupress.indiana.edu

Telephone orders   800-842-6796
Fax orders   812-855-7931
Orders by e-mail   iuporder@indiana.edu

© 2007 by The Ogden Union Station Foundation and Ric Morgan
All rights reserved

No part of this book may be reproduced or utilized in any form or by any means, electronic or mechanical, including photocopying and recording, or by any information storage and retrieval system, without permission in writing from the publisher. The Association of American University Presses' Resolution on Permissions constitutes the only exception to this prohibition.

The paper used in this publication meets the minimum requirements of American National Standard for Information Sciences—Permanence of Paper for Printed Library Materials, ANSI Z39.48-1984.

Manufactured in the United States of America

Library of Congress Cataloging-in-Publication Data

Morgan, Ric.
   The Train of Tomorrow / Ric Morgan.
      p. cm. — (Railroads past and present)
   Includes bibliographical references.
   ISBN-13: 978-0-253-34842-5 (cloth : alk. paper)
   ISBN-10: 0-253-34842-0 (cloth : alk. paper)   1. Train of Tomorrow (Passenger train)—History.   2. Dome cars (Railroads)—United States—Design and construction—History—20th century.   3. General Motors Corporation—Research—History—20th century.   I. Title.
   TF456.M67 2007
   625.2'3—dc22
                                                2006030068

1 2 3 4 5 12 11 10 09 08 07

*To Ruthie and Pop
who, through their love and support,
encouraged me to do the best
at whatever I chose to do and
introduced me to trains.*

CONTENTS

Foreword by Roger Smith   ix
Preface   xi
Acknowledgments   xv

Part 1. The History of the *Train Of Tomorrow*

1. "If They Could See What I See . . ."   3
   The First Dome Cars   6
2. The $100,000 Model Train   8
   Early Design Ideas   13
3. Excuse the Dust: Astra Dome under Construction   19
4. Something for Everyone to See   28
   News of the Day   34
   The *Blue Goose*   38
   The Chicago Railroad Fair   50
5. For Sale: One Train   54
6. Going to Work   60
7. Life after Death   66
   Ogden Union Station   74

Part 2. The *Train Of Tomorrow* Inside and Out

8. Locomotive 765   81
9. *Star Dust*   86
10. *Sky View*   97
11. *Dream Cloud*   115
12. *Moon Glow*   127

Appendix A. The Diary   139
Appendix B. Technical Information   177

Glossary   203
Notes   205
Bibliography   209

FOREWORD

General Motors Corporation has had a long and proud tradition in the railroad industry. Its history has its roots in innovation. It began over 80 years ago by revolutionizing the railroads with the introduction of the 2-cycle diesel engine. GM has produced well over 50,000 diesel-electric locomotives and a majority of the locomotives in service in North America. In over 100 countries around the world, GM locomotives power the railroads, often at the far reaches of civilization.

But GM is more than a manufacturer. It is a researcher and developer, always applying its ingenuity to push the limits of modern technology. They put electronics, on-board computers, and sophisticated control systems into today's locomotives. The future will no doubt see even more exciting progress.

The *Train Of Tomorrow* is an excellent example of GM's role in the evolution of the railroad industry, and it shows the same emphasis on passenger comfort that's evident in GM's automotive products. I'm pleased that its story is now being told in such comprehensive form. This book represents an opportunity to support and preserve part of our great nation's past—the railroad industry.

Roger Smith
Former Chairman and CEO
General Motors Corporation

PREFACE

I AM IN LOVE, AND I HAVE BEEN EVER SINCE I FIRST SAW A PICTURE OF THE *Train Of Tomorrow*. I don't remember where or when it was, but I knew I had to know more about her.

I am not a dyed-in-the-wool rail fan. Oh, I enjoy riding trains and listening to railroad business on a scanner, but I do not own a lot of railroad memorabilia, nor do I go out and hunt down railroad prey. Trains are my favorite form of transportation; I am always ready to hit the rails, but I do not spend every moment of my life thinking, eating, or sleeping trains.

However, I am in love with the *Train Of Tomorrow*. I am not sure what it is about the train that's so appealing to me. Maybe it's the lines of her construction: she looks so sleek and elegant yet sturdy. Maybe it's the uniqueness of the concept: four dome cars designed to provide a smooth, noiseless ride in unequaled luxury. Perhaps it's the excitement the train generated when it was on tour and again in revenue service. Whatever it is, it's a certainty that I am obsessed with knowing all I can about the *Train Of Tomorrow*.

I remember seeing a picture of the train in one of the leading railroad magazines. Being instantly hooked, I wanted to know more, so I went looking for any other information I could find. There was little available. I couldn't believe that something this wonderful could go unnoticed. Even a trip to the library turned up very little. Then I struck gold—or so I thought at the time. Patrick Dorin had included a chapter about the *Train Of Tomorrow* in *The Domeliners: A Pictorial History of the Penthouse Trains*. But it was not enough; I wanted to know more.

A couple of years after a move to Detroit, I worked up the courage to make contact with the corporate library at General Motors. After establishing that I really had an interest in the subject, the librarians invited me to visit. When I arrived for my appointment, the librarians laid before me a mother lode of information about the train. I will always be indebted to those lovely ladies for their kindness.

After digging through all the information, I still was not satisfied that I knew all there was to know. My next move was to look for a book about the *Train Of Tomorrow*, but none existed. It was then that I decided to write that book. Insanity does not run in my family, but sometimes when I am struck by an idea like this one, I begin to wonder. Little did I know then of the adventure I was in for over the next six years.

Doing a historical nonfiction book is a lot like being a detective trying to solve a crime. Sometimes, in doing the research, I would come to a dead end or discover I was going in

The picture that started it all.
GM.

the wrong direction. However, usually someone there helped me get back on the right track.

Since the *Train Of Tomorrow* was a one-of-a-kind project, it was constructed with meticulous care by Pullman-Standard; it's evident that the train was one of the best thought-out, designed, and constructed trains ever built. Throughout all of the work done to produce this book, I never heard one person say anything bad about the train. People whom I met who worked on the train have the fondest memories of it and will often say it was one of the best experiences of their lives. There are very few people still alive who worked on or around the train, although I have met many people who remember seeing it, or even hearing about it, when it was a current topic. What has been interesting is the renewed interest in the train. Everyone whom I have told about the train is fascinated with it. I hope that this book will help people learn about one of the most important trains in railroading history.

When you do the research on something like this, you begin to question its place in history and its social significance. Like many train fans, I have my fond memories. Two involved dome cars on the original auto-train™. The first memory is of having dinner and breakfast in the dome of the dining car. Watching the sunset and sunrise while eating a delicious meal moving along at nearly 80 mph is something you never forget. Following dinner, my friends and I moved to the full-domed lounge car. A thunderstorm raged outside while we enjoyed drinks and listened to an entertainer sing and play the guitar. Rain pounded the glass above as we watched lightning strike all around us. There aren't many places you can do that other than in a dome car.

The *Train Of Tomorrow*'s place in history is imbedded in those memories of a shared experience with fellow passengers. From the shores of the Maritimes in eastern Canada, to the magnificence of the Rockies, sitting in a dome car looking at the ocean or snow-capped mountain peaks heightens the experience of each place. The *Train Of Tomorrow* inspired car builders and railroads to enrich the train travel experience. Always a luxury, dome cars brought unique perspective to a broad continent of exciting and moving scenery.

Social significance is rather hard to define. The *Train Of Tomorrow* was innovative in many ways besides the domes—the extensive use of mobile electrical power; outside swing hangers on the trucks; the use of new building materials; creative modern décor; and mobile communications. In a very generous move, General Motors decided not to patent the dome design, allowing anyone to use it. These ideas were put to use by all the car builders to make rail travel safer and more comfortable. These innovations and others helped revitalize rail travel following World War II, even though these efforts didn't help save passenger service for very long.

Why would General Motors want to build the *Train Of Tomorrow*, when they were in the business of building locomotives through their Electro-Motive Division? As GM president C. E. Wilson explained at its inauguration on May 28, 1947, "The *Train Of Tomorrow* is an experimental project to try out ideas for improvement of railroad travel in future years. General Motors has no intention of going into the passenger car business. The *Train Of Tomorrow* is a research project . . . in which a motive-power manufacturer, the railroad car builders, and the great railway systems of America have pooled their accumulated skills. We thought this was a good demonstration of what could be done by cooperative action of many men, many engineers with a common idea of making something that was better for their purpose."

Richard Terrell, who worked on the train while it was on tour, said, "It was a show train, I don't think you could say, 'We sold locomotives because of this.' But I think it helped our credibility a great deal with the railroads because, remember, we were really the young upstarts, the innovators, prior to World War II. We still weren't accepted fully as dedicated railroad suppliers, and one of the things that the train did was show railroads a willingness to spend our money and our effort to help them stimulate their image." According to its public relations manager, Eddie A. Braken, the *Train Of Tomorrow* was "one of the great goodwill things in that era, probably one of the very best of the things we did to say GM is a great company and 'see what we can do.'"

It is my hope that you will grow to love the *Train Of Tomorrow* as much as I have. Please know how much I have enjoyed working on this project. It wasn't always easy, but it was well worth the effort and the adventure. I learned almost everything there is to know about the train, and I met some terrific people: generous, kind, understanding, and even willing to bail me out of jail (don't ask). Many of these people have become friends. I have also seen and done a lot of things I am sure many people would envy.

There is one person who must be mentioned here rather than in the acknowledgments. A big debt of gratitude goes out to Cyrus R. Osborn, creator of the *Train Of Tomorrow*. His idea and persistence made a dream of a train with "glass roofs" a reality. Without him, there would be no *Train Of Tomorrow* and no book about it.

I am in love with the *Train Of Tomorrow*, and I hope that after reading this book, you will be too.

ACKNOWLEDGMENTS

IT IS NEARLY IMPOSSIBLE TO DO A BOOK LIKE THIS WITHOUT THE HELP OF A LOT OF people. From individuals to corporate employees, over 50 people have contributed to this book in one way or another, from major research to making suggestions for improvement. Many went out of their way to do more than anyone could expect.

First, to Mike Burdett and Lee Witten, two gentlemen whose efforts made the publication of this book possible. Without them and their belief in the book, the manuscript would still be sitting in a file cabinet.

Next, my editor, Linda Oblack. Her calm manner and gentle voice helped in moments of distress. She used a firm but gentle hand in guiding me through the whole process and was helpful in making suggestions, yet she gave me enough credit as an author to produce a publishable book by doing what I thought was right for the benefit of the reader of this historical tale.

I thank Jane Ronca, a special lady who did the proofreading and made excellent suggestions on how the book could be improved. Rosalind Payne-Thompson and DeCaroll Baker made valuable contributions by helping with some of the proofreading chores. Their comments were also welcomed. Through the generosity of her time and use of her skills, Margaret Czarniak made the typing load much lighter. And a special thanks goes to Phil Cooper for his encouragement and insight.

Otto Masek and Tom Cashmer of Pullman Technologies bailed me out when I got into trouble, and Joe Krajniak and Nancy Calvert from EMD helped in innumerable ways to ensure that I received some of the very important things necessary to make this book a reality. Thank you all for being there when I needed you.

Finally, I am grateful to a very special group of people who did everything they could to make the research for this book easier. Many of them opened their doors and files to me. A "thank you" is not enough for all they have done, but they will know that it is from the heart. Perhaps the best thing to do for them is for me to write and publish a good book: Wallace W. Abbey, Virginia Abel, Elwyn Ahnquist, Mary Arnold, James Benak, Eddie Bracken, Edward A. Braken Jr., Bill Brennan, E. Preston Calvert, Jim Couglin, Fletcher Cox, Kevin Dabe, Sid Dawson, Mike Deja, Helen Donovan, Henry Fernandez, Bob Geier, Mike Giefer, Teddy Griffith, Edmund P. Hogan, Howard Hopkins, Harvey Jackson, Eugene J. Jendrasik, Pamela Johnson, Charles E. Kinzer, Bill Knick, Don Kratzer, Dan Kuhn, David Lawrence Jr., Terry Lieb, James Limbacher, Richard Luckin, Art Lloyd, Ken Longe, Sandra Marshalek, Bill McCarty, Renee McHenry, Ken Meeker, Rudy Morgenfruh, Ken Oyer, Alan Perlman, Ed

Pohlman, Don Postma, Kathleen Presnail, Dick Ryker, Robert Shaw, Martin B. Shellenberger, Jim Shields, Terri Sinnot, Roger Smith, Don Snoddy, Sandy Spikes, P.W. Stafford, Carol J. Stetter, Neal Strodel, George Swallow, Lester Tippe, Hugh Wells, Mike Wenninger, Mr. and Mrs. Walt Whitmire, Nora Wilson, Dale Wood, Gene Wright, members of the Railway and Locomotive Society, and members of the Union Pacific Historical Society.

Trademarks and Copyrights

The Ogden Union Station Foundation has acquired the rights to the names *Train Of Tomorrow, Astra Dome, Astra Liners, Star Dust, Sky View, Dream Cloud,* and *Moon Glow.* The *Train Of Tomorrow* logo, photographs, and documents concerning the *Train Of Tomorrow* are trademarks, service marks, and/or the property of the Ogden Union Station Foundation. This includes the trademarks, logos, and copyrights to all materials, drawings, photographs, films, and video reproductions.

General Motors owns the trademarks and copyrights to General Motors Corporation, General Motors, GM, EMD, Electro-Motive, Delco, Frigidaire, Hyatt Bearings, Detroit Diesel, and all other GM divisions and products connected.

Union Pacific, Union Pacific Railroad, UP, UPRR, the Union Pacific shield, and photographs and documents concerning the *Train Of Tomorrow* are trademarks, service marks, and/or the property of the Union Pacific Railroad Company, Omaha, Nebraska.

Pullman and the photographs, drawings, specifications, and documents concerning the *Train Of Tomorrow* are trademarks and/or the property of Pullman Technologies, Harvey, Illinois. Pullman Technologies is in no way connected with Pullman-Peabody or Trinity Industries.

Burlington Northern and the Burlington Northern photographs of the dome car *Silver Dome* are trademarks and/or property of the Burlington Northern Railroad and appear with oral and written permission of the Burlington Northern Railroad.

Canadian Pacific, CP Rail, and the CPR photographs of the mountain observatory cars are trademarks and/or property of the Canadian Pacific Railroad and appear with the written permission of the Canadian Pacific Railroad. All rights reserved.

The photocartoon in chapter 10 is the property of Kalmbach Publishing Company and appears with written permission. All reproduction rights are reserved.

The E7 art work on the January 1979 cover of *Trains Magazine* in chapter 8 is the property of the Kalmbach Publishing Company and appears with written permission. All reproduction rights reserved.

The September 1948 cover of *Railroad Magazine* in chapter 4 is the property of Carstens Publications, Inc., and appears with written permission.

Photographs from the Wallace W. Abbey Collection are the property of Wallace W. Abbey and Pinon Productions and appear with written permission. All other rights reserved.

Photographs from the Richard Luckin Collection are the property of Richard Luckin and appear with written permission. All rights reserved.

Photographs from the David Seidel Collection are the property of David Seidel and appear with written permission. All rights reserved.

Photographs from the Rudy Morgenfruh Collection are the property of Rudy Morgenfruh and appear with written permission. All rights reserved.

Photographs from the Lee Witten Collection are the property of Lee Witten and appear with permission. All rights reserved.

Photographs from the Harry Stegmaier Collection are the property of Harry Stegmaier and appear with permission. All rights reserved.

Libbey Glass and the Libbey Glass photographs in appendix B of the glassware used on the *Train Of Tomorrow* are the trademark and/or property of Libbey Glass, Inc., a subsidiary of Owens-Illinois, and appear with the written permission of Libbey Glass, Inc.

THE TRAIN OF TOMORROW

PART ONE

THE HISTORY OF THE
Train Of Tomorrow

The world has always been full of dreamers. But only a few see their ideas become reality. Cyrus R. Osborn was such a dreamer. His idea for a glass-enclosed room on top of passenger cars captured the imagination of many of his peers, who did all they could to see it become the *Train Of Tomorrow*. Many dreams take years or even decades to accomplish, but the *Train Of Tomorrow* went from idea to reality in less than three years.

Although the life of the *Train Of Tomorrow* was short, it was eventful. It was widely received by millions of people, written about in nearly every newspaper and magazine in America, swooned over by celebrities, given numerous honors, and praised as being ahead of its time. As a "rolling laboratory" designed to promote rail travel, it helped to rekindle the desire for train travel in a nation that had been denied travel for pleasure during World War II. As a creative endeavor, the train showcased General Motors railroad-related products as well as being an example of what private industry can do in a free enterprise system.

The history presented here is the result of six years of digging through files, photographs, drawings, and blueprints and pulling together thousands of facts and figures. Since the seed of the idea was germinated almost 60 years ago, records have been lost. For the locomotive and three of the cars, life on the rails lasted only 14 to 18 years. For one car, life goes on.

This section deals with the history, development, construction, and life in service of the *Train Of Tomorrow* and the people who were intimately involved with it.

CHAPTER 1

"IF THEY COULD SEE WHAT I SEE…"

The story goes that in July 1944, Cyrus R. Osborn, general manager of the General Motors Electro-Motive Division (EMD), was making an inspection tour of wartime freight movements on-board the Denver & Rio Grande Western Railroad. Riding in the fireman's seat in the cab of a diesel locomotive, he commented to everyone present that "a lot of people would pay $500 for the fireman's seat from Chicago to San Francisco if they knew what they could see from it."[1] That story was printed in hundreds of newspapers and was repeated in speeches and radio programs across the country.

Well, that's a nice story, but it is not entirely accurate. At the inaugural luncheon for the train in Chicago on May 28, 1947, Osborn explained how the idea came about for a car with either a separate glassed-in room on top or a glass roof:

> Back in July 1944, I happened to be making one of my customary trips over the railroads and, in this case, I had made up my mind to ride a freight locomotive from Denver clear through to California. We left Denver early in the morning on a perfectly gorgeous day. Within a short time, we were in the midst of the Rockies. While I had traveled over this route many times, I had never realized nor been able to enjoy and appreciate the beautiful scenery along the route of the Denver and Rio Grande Western Railroad. I made this remark to one of the railroad officials who accompanied me, and he replied that, while he had traveled over his railroad hundreds of times, over many years, he had never really known the railroad until he had made the trip in the cab of a diesel locomotive, where it was possible to have a clear unobstructed vision.
>
> This conversation started a train of thought and discussion with one of our own people, Ernie Kuehn. We discussed the adaptation of a flat car which years ago operated through the Royal Gorge where, as a result of exposure to smoke and dirt, you could only see the whites of the eyes of your neighbor when you finished the trip. We discussed car construction made entirely of glass from the belt line of the car over the roof; for safety reasons this did not appear practical. We then became preoccupied with other matters, and this was the extent of our discussion this first day.
>
> We got off the train at Grand Junction, and on the following day we again started out early. We rode a diesel freight locomotive to Soldier Summit. At this point, since the rest of the trip was downhill, they took off the diesel locomotive and put on a steam locomotive. At this time, I went back to the rear of the train and rode the caboose into

Cyrus R. Osborn. EMD.

Salt Lake City, which was a tremendously interesting and enjoyable trip. It was really the ride in this cupola of the caboose that gave me the idea.

When we arrived at Salt Lake City, I remarked to Ernie Kuehn that if the traveling public only knew what they could see from the operating cab of a diesel locomotive or caboose of a freight train, the railroads could sell seats in these two places at $500 apiece and always keep them full. When we arrived at Salt Lake City late in the evening, we went to the hotel and spent several hours of the night sketching up the idea which later became the Astra Dome.[2]

After returning home, Osborn presented four questions to the EMD engineers:

1. With the use of steel ribs and the heavy, bullet-proof glass developed during the war, could a dome be built that would provide the necessary passenger safety?
2. Would it be low enough to clear most tunnels, bridges, and station roofs in the United States?

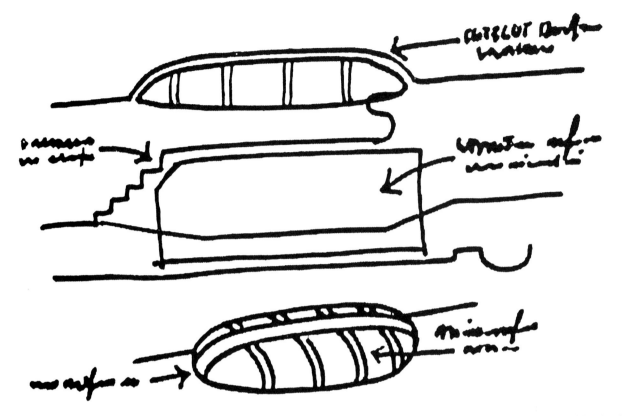

3. Could the regular floor of the car be depressed sufficiently to provide headroom beneath the dome, as well as in the dome, and still meet clearance requirements of the railroads?
4. If the regular floor line were thus altered, could the buff and drag of moving cars be transmitted through a redesigned underframe instead of through the conventional heavy center sill?[3]

A reproduction of the sketches made by C. R. Osborn in July 1944 of his idea for a domed passenger car, drawn on stationery from Hotel Utah in Salt Lake City. GM.

Osborn asked his engineers if his idea was practicable. They reminded him, "There must be sufficient room within bridge, tunnel, and station limits because cupolas of cabooses, which rise considerably higher than standard passenger car roof lines, go through."[4] The engineers said the car should meet Interstate Commerce Commission requirements for car design and safety and be able to accommodate the mechanical components necessary to operate a modern railroad car, yet leave the space below the dome for passenger use.

It wasn't long before the engineers decided that not only was it possible but changes in design necessary to accommodate the dome and the lowered floor in the area under the dome would exceed minimum standards for safety.

From EMD the idea was passed to Harley Earl at GM's Styling Section in Detroit. Earl decided "to build a small model incorporating the principal idea of the Astra Dome, and secondly to do a complete study of the things which the traveling public liked about existing railroad cars and, more importantly, the improvements they would like to have."[5]

THE FIRST DOME CARS

Cyrus Osborn's "penthouse on rails" is an idea that dates back to the nineteenth century.

T. B. Watson, a conductor for the Chicago & North Western Railroad, was the first to suggest adding a second story onto a railroad car. Assigned to a run through Iowa in 1863, Watson arranged a pile of boxes under a hole in the roof of the caboose and seated himself on top of the boxes with his head sticking out through the hole. According to Donald Dale Jackson, Watson so enjoyed the view that he persuaded C&NW officials to add cupolas to cabooses under construction.

An 1891 issue of *Scientific American* reported that T. J. McBride of Winnipeg, Manitoba, had patented and constructed a sleeper with three domes.

At about the same time, R. C. Riblet, the chief engineer of the Coeur d'Alene Railway & Navigation Company, designed an electric dome car for transit use. The 40-foot, 60-passenger car had a luxuriously appointed dome section that accommodated both the passengers and the motorman, with the area below the raised floor of the dome section reserved for freight. It is not known whether the car was ever built.

In 1902, the Canadian Pacific Railroad built an observation dome car based on McBride's concept in the Hochelaga shops in Montreal. Originally numbered 517, the car had two observatories that looked like enlarged caboose cupolas that extended 3 feet above the roofline. Weighing more than 30 tons, the 64-foot car (over the couplers) had a natural mahogany exterior finish. The capacity was 50 passengers, with seats for 12 in the two observatories.

In 1906, CPR built three more such cars in their Angus shops, also in Montreal. Numbered 84, 85, and 86, the new cars had large windows as well as a glass roof between the two cupolas or "domes." The glass in the center of the car had large roll-type blinds to keep out the heat, but evidently they didn't do a very good job, as heat-resistant and glare-resistant glass materials (Thermopane) had not yet been invented. About the same time, No. 517 was renumbered 83 and a glass roof was added. The original car had swiveling leather chairs in the center section, with cane-bottomed swivel seats in the dome section. The three later versions of the car had bench seats running down the center aisle that faced the large windows.

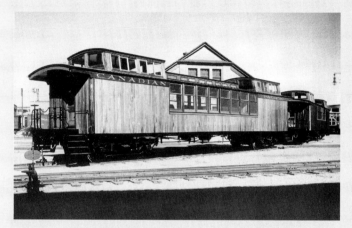

The second generation of Canadian Pacific Railroad mountain observation cars built in 1906. CP Rail.

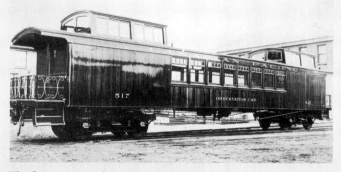

The first mountain observation car, 517, built by the Canadian Pacific Railroad in 1902. CP Rail.

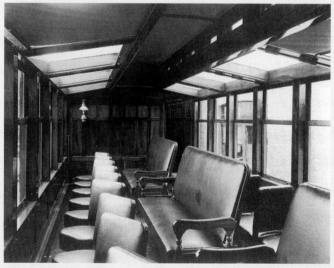

The interior of mountain observation car 84. CP Rail.

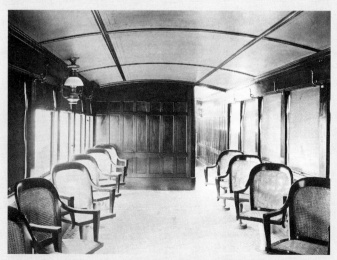

The interior of mountain observation car 517. CP Rail.

Coupled among the sleeping cars for use by the first class passengers, the observatory cars were used during the summer months in British Columbia between Sicamous and Revelstoke. At first the cars were very popular, but because of heat build-up from the large windows in the center section and the glass roofs, the cars were dropped from passenger service in 1909 and replaced by a conventional sleeper-buffet-lounge car with large observation platforms surrounded by decorative brass railings. After retirement from passenger service, the cars were used as service cars for repair and

THE FIRST DOME CARS continued

maintenance crews and then scrapped in the late 1910s and early 1920s.

Records at Union Pacific mention a "dome" car having been built in the 1930s by the Wichita Falls and Southern Railroad in Texas, later a part of the Rock Island system. The records state that the car was a combination parlor-coach-baggage-observatory-dome car and that a photograph of the car was made in 1939. In fact, the dome section was very much like the observatories on the Canadian Pacific cars: a modified cupola.

In spite of the long history of the dome observatory car, Cyrus Osborn is still considered the "father" of the modern dome car. To commemorate the birth of the dome car, the Denver & Rio Grande Western Railroad placed a monument in Glenwood Canyon, Colorado, in 1950. The monument had a two-legged stone pedestal for a model of a dome car. A plaque read:

> The idea for the Vista-Dome railroad car was conceived on the Denver & Rio Grande Western across the Colorado River from this point on July 4, 1944. Riding through Glenwood Canyon in the fireman's seat high in the nose of a Rio Grande diesel locomotive built by his company, C. R. Osborn, vice president of General Motors and general manager of Electro-Motive Division, was struck by the need for giving passengers an unobstructed view of the inspiring scenery overhead and on all sides. The idea of building glass-enclosed domes into the cars occurred to him. Unlike so many originators of unusual ideas, Osborn in a brief five years saw his dream grow into full practical utilization. Vista-Dome *California Zephyr* trains went into service March 21, 1949, between San Francisco and Chicago and now pass the spot where the idea was born.

In 1985 the monument and model dome car were sent to the Colorado Railroad Museum, where they were placed on public exhibit.

At this point, there is a conflict in the story. One version goes that one person, George Jergenson, was assigned to ride trains of every description for several months in order to gather information on current car design and what passengers would like to see changed. The other version is that nine designers from the Styling Section traveled around the country on all kinds of trains for two months, even going so far as being allowed to take their wives and children along. Probably a combination of the two versions is correct. Harley Earl sent a letter to Osborn on October 3, 1944, in which he said George Jergenson was to report for work that afternoon and that five designers and some sculptors were to be put on the project. Earl went on to say that the designers who had worked at some of the motion picture art studios had just arrived from California. "We are hitting this with a bang and will do it as fast as possible." Earl said the project, including art work, designs, and model, would take about 90 days and would cost about $25,000.

In either case, the designers came back with a long list of ideas, which they translated into 1,500 sketches in crayon, watercolor, or pencil. Again, the only limitations were "a few obvious engineering requirements, such as permissible length and height, and allowance for heating, air-conditioning, and air brake equipment."[6] After evaluation by the engineers at EMD, 100 of the design ideas survived. As stated in the GM press handout, "Mr. Osborn had presented merely an idea. Mr. Earl and his staff had developed it into a practical and beautiful actuality."

CHAPTER 2

THE $100,000 MODEL TRAIN

Along with the sketches of domed railroad cars made by the Styling Section of General Motors, a small model of one car was built to get railroad executives' reactions. However, the model was deemed too small for public viewing, and it was suggested that it might be unwise to have only one domed car on the train because of the limited number of seats available in the dome compared with the total number of people on most trains at the time. So the decision was made to develop designs for all types of cars with domes and to build a bigger model. Design and construction for a chair car, sleeper, diner, and club car began. It was planned that once the model was finished, it would be put on display for the executives of the Class 1 railroads for evaluation and feedback on the dome idea. In the event the reaction was negative, GM was prepared to drop any further development of the dome. One important point was stated throughout the project: GM did not plan to become a builder of railroad passenger equipment. According to a GM public relations handout, "This study is presented to the railroads of America as a supplement to their own forward thinking and as a contribution by Electro-Motive Division to the industry with which it is so completely involved." In fact, GM was getting involved only because the idea had come from one of its executives, vice president Cyrus Osborn.

The sculptors and model builders of the Styling Section began working on a design that would be 45 feet in length. Made of wood, plastic, and metal, each car measured 10 feet, while the model of the locomotive, added later, was 5 feet long. The interiors were populated with 175 painted clay figures of men, women, and children in sitting, standing, and walking poses just like passengers on a moving train. Each figure was "dressed" appropriately for its location in the model. In a drawing room in the sleeping car, for instance, a nude clay figure stood in the shower. This was scandalous for 1945! When the scale model of the locomotive had been added, there were even two clay figures wearing white uniforms at the controls in the cab. Interior furnishings were complete with tables, chairs, and berths as well as kitchen and bar equipment. The walls were painted solid colors or in such a way as to imitate wallpaper. Some walls were covered with mirrors. Nothing was left out. Three-quarter-inch knives and forks made from aluminum sheet were laid on the tables in the dining car along with plastic plates, cups, and glasses. The kitchen was supplied with food. Tiny glasses and bottles had been made from rods of Lucite for the cocktail lounges. Replicas of popular magazines and tiny packs of cigarettes were placed casually on tables throughout the train. An intricate lighting system illumi-

Model builders carving the model of the locomotive. GM.

Some of the clay figures and furnishings for the model. GM.

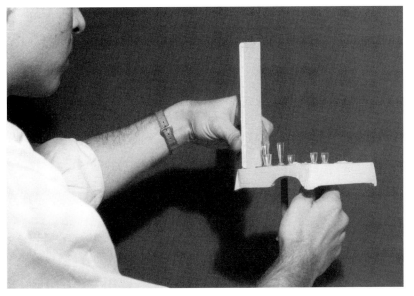

Showing the scale of a table setting for the dining car. GM.

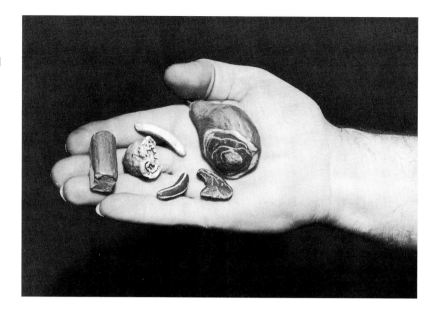

Miniature food items made for the model kitchen: a loaf of bread, a head of lettuce, a banana, a zucchini, a steak, and a ham. GM.

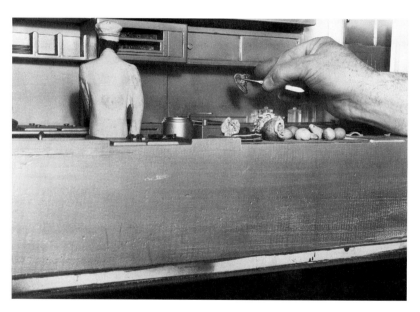

Placing a miniature steak in the kitchen of the model. GM.

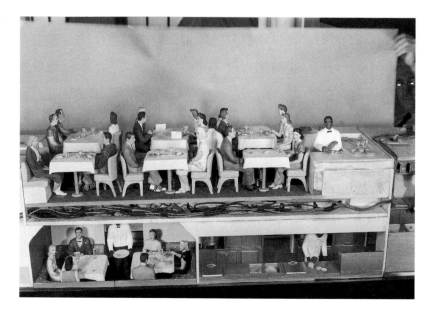

The interior of the dome dining room, private dining room, and kitchen of the dining car *Sky View*. GM.

Placing the body of the dining car *Sky View* on the model. GM.

A couple getting ready for bed in a compartment on *Dream Cloud*. GM.

A drawing room setting on *Dream Cloud*. Note the nude clay figure of a woman taking a shower at the left. GM.

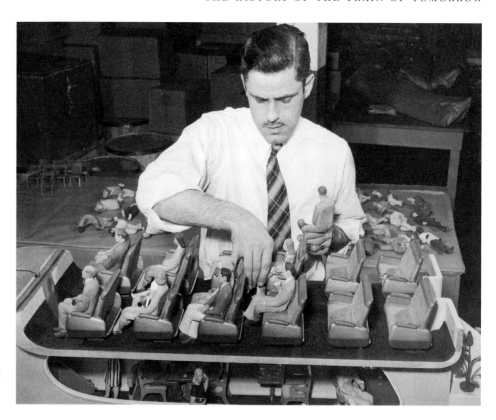

Placing some of the 175 clay figures of the train's model in the dome of the observation lounge car. GM.

nated even the smallest detail of the interiors. More than 1,100 objects were created for the dining car alone.

As construction continued, plans were being made for the exhibition site. Permission was granted by a GM division office in Chicago for use of a former Cadillac showroom at the corner of Washington and Kenilworth in Oak Park, Illinois, that had been closed during the war. A display was built using the sketches and drawings of the 100 ideas that had been approved by the engineering staff at EMD from the 1,500 ideas submitted early in the project. The backdrop for the model was a western mountain scene, painted and hung on one wall of the showroom. A miniature track was built on a wooden scaffold fronted by a rocky ledge. At the both ends of the track, tunnel openings were built to add to the realism of the display. Most of the parts of the display were moveable so they could be rearranged as needed. Stage lighting was installed, and then the model was brought in from Detroit by truck and set up on the track.

The first showing was made to Cyrus Osborn's close friend Ralph Budd, president of the Chicago, Burlington & Quincy Railroad, and to other officials of the railroad. The reaction was so strongly in favor of the concept that Budd didn't waste much time ordering one of Burlington's coaches in the shop for repairs to be converted to a dome car. No. 4714, *Silver Alchemy*, was converted and renamed *Silver Dome*. According to a GMC public relations handout, "The depressed floor under the dome was omitted because materials and time were not available and because the principal idea was to test public reaction to the ride in the dome." In June 1945, *Silver Dome* was taken on a tour of the CB&Q system and was later put into service on various routes. At first, passengers were allowed to ride in the dome for about 30 minutes and then asked to complete a survey form. Over 10,000 people wrote comments about the ride in *Silver Dome*, some of which were reprinted in a GMC brochure, *The Essence of Progress*:

> It is lovely-cool-comfortable and roomy—makes traveling something to look forward to.
> A large movement in postwar pleasure.

EARLY DESIGN IDEAS

Nine designers of the GM Styling Section submitted over 1,500 ideas, 100 of which were approved. Some of the early design ideas were then refined or so transformed that they were no longer recognizable. A few were incorporated into the original model but were later labeled impractical and changed. The name of the train serves as the first example: it was supposed to be referred to as an *Astra Liner* instead of the *Train Of Tomorrow*. According to a July 1946 EMD press release, other ideas that were later modified included additional lavatories in the space below the dome in the coach, a large women's lounge and children's padded playroom, and an infant-care room with a built-in changing table, high chair, bottle warmer, a refrigerator for formula, and a sink with a high faucet, as well as a card room, a snack bar, a smoking room, and family rooms. This last idea was actually used on the *Train Of Tomorrow*: three separate seating areas were divided off to be used exclusively for families or small groups of people traveling together.

The original concept of the dining car was that it would be of the same design as a conventional car, without the dome but with a glass roof over the main dining area. Plastic shades were to be used to keep out undesirable sunlight. There was also to be a cocktail lounge waiting area at the end of the car opposite the diner. Several of the tables were designed to fold into the wall, allowing couples to dance the night away under the stars.

A model car of this design was built and displayed, but it gave way to another early idea for a dome car that was to have dining rooms at both ends, with the kitchen and pantry located under the dome. In fact, much of this design did appear in the construction of the dome diner, *Sky View*, but with certain modifications.

Ideas for the dome sleeping car included couches with loose cushions and luggage storage spaces; dressing tables; walls with scenic panels, recessed lighting, and compartments beside the beds for water carafes, books, purses, etc.; folding screens surrounding the toilet area; showers; and a suite with a double bed.

Design suggestions for the observation lounge car that never made it past the concept stage include putting chaise longues in the dome for sunbathing or napping on transcontinental routes.

Even though most of these ideas didn't make it into the final construction plans for the *Train Of Tomorrow*, the train was still far ahead of its time in many ways, and several of the ideas are still used on passenger cars today.

A traveler's dream by rail realized.

Better view of scenery especially for long-distance travel. I prefer it to the airplane.

I feel like I'm riding in an airplane, and the view is beautiful and resting.

Wonderful view, solid comfort even from sun rays, freedom from wheel and track noises.

The pleasantest ride I've ever had. Promises are great for future traveling. Congratulations!

It's the best idea yet to enable you to see where you're going and would be wonderful on a moonlight nite.

A short time later, CB&Q built a second dome car, *Silver Castle*, and put it into revenue service.

In spite of being beaten out by others as the first dome car to be built, the *Train Of Tomorrow* did retain a couple of legitimate firsts. It had the first dome coach based on new construction and not as a remanufactured car. It also had the first dome sleeping car and the first dome diner.[1] Only 235 dome cars were ever built.

Between February 21 and June 23, 1945, the exhibit was shown to more than 350 railroad officials from 55 Class 1 railroads as well as to representatives of the railroad car-building companies. Many suggested changes. Suggestions were scrupulously recorded by GM staff, and many were utilized in the final design.[2]

Two models were built. The first had only three cars: dining, sleeping, and observation lounge. The dining car had a glass roof but not a dome. The sleeper and observation car had domes with curved glass. The rear of the observation car was built in such a way that many people, upon seeing a photograph of the model, have mistaken it for the front of the locomotive. It's assumed this was the model the railroad executives saw in Oak Park. The second model proved to be a more accurate representation of how the train would look when it was built by Pullman-Standard. There is no record of when the transition be-

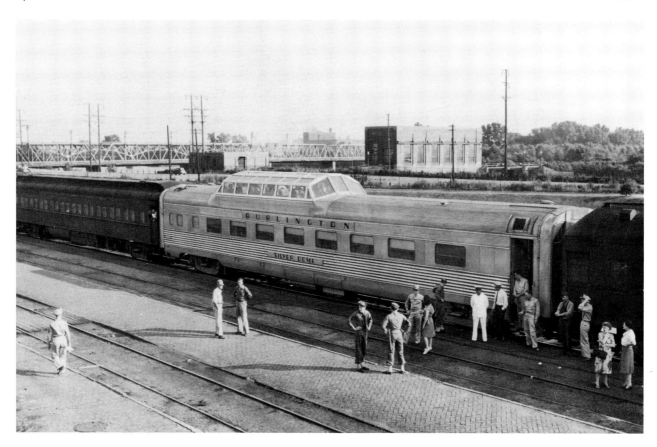

Burlington's *Silver Dome*, the first modern dome car. During a test run on July 27, 1945, the Vista-Dome is shown at a stop at Ottumwa, Iowa. Burlington Northern Railroad.

tween the two models took place, but based on a November 7, 1945, letter to Osborn at EMD from Harley Earl at the GM Styling Section, there is some indication that the change in design occurred some time after the contract was made with Pullman-Standard. Earl wrote: "I will make provisions . . . in my budget . . . for any additional expenses of bringing the model up-to-date with the production agreement." Obviously, the interior of the dining car had to be changed with the addition of the dome, but close examination of photographs of the second model reveal that not much else was changed. To add to the confusion, one of the photographs of the second model shows five dome cars being displayed.

While the original appropriation for the model in 1944 was for $25,000, expenses had risen to $101,772 by November 1945. In his letter to Osborn, Earl wrote, "This amount represents direct charges only. I feel the costs are excessive due to the necessity of training personnel and gathering the necessary know-how of approaching this kind of a project which certainly would not be true if we were to take on such an assignment in the future."

According to a GMC promotional pamphlet, *The Essence of Progress,* "A number of railroads contracted for dome cars, each of the car building companies receiving orders. The car builders were given letters of permission to use the GM designs without fee." Consequently, 49 dome cars were on order with the Budd Company by the time the *Train Of Tomorrow* made its debut.[3] There was no problem with others copying the idea, because GM saw the program not only as a research project to stimulate interest in new car design and construction but also as a way to encourage railroads to order new cars with GM railroad-related products on-board. GM made the idea available to any car builder or railroad that might be interested at no cost for licensing or royalties.

Life for the model did not end there. The revised model was put on display repeatedly. Usually it would be seen at auto shows and other GM exhibitions up through 1949, even though the *Train Of Tomorrow* had been built and was on tour. The model was displayed

Three cars of the original model. Note how the shape of the last car gave a lot of people the impression that this was the front of the locomotive. GM.

This close-up of the dining car shows how the designers originally planned a glass roof instead of a dome. GM.

The second model of the *Train Of Tomorrow* on display in a former Cadillac dealership in Oak Park, Illinois. GM.

The second model on its display stand in Oak Park, Illinois. GM.

The second model from the rear. Note that there are five domes in this shot. GM.

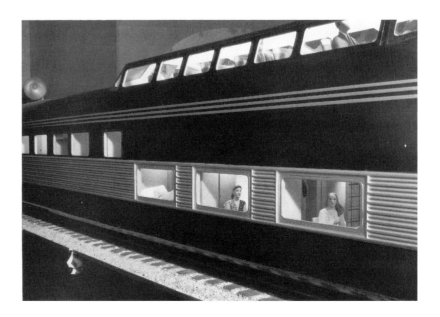

Exterior of the sleeper *Dream Cloud* in the second model. GM.

16

Wow! GM.

Executives inspect the model train. GM.

The model of the *Train Of Tomorrow* on display at the Waldorf=Astoria for the 1949 GM New York Auto Show. GM.

17

at the Chicago Museum of Science and Industry in the Motorama exhibit from April 20, 1949, to November 7, 1955. The last known place the model was displayed was at an award ceremony honoring Cyrus Osborn in Cincinnati in 1955. It isn't known what happened to the model after that. It is generally known there are some people at GM called "squirrels" who store things that are slated for disposal. Quite a few people at GM believe the model is in the possession of one of these. If this is true, the question arises as to who that person might be.

CHAPTER 3

EXCUSE THE DUST: ASTRA DOME UNDER CONSTRUCTION

After 350 railroad executives from 55 Class 1 railroads saw the model of the *Train Of Tomorrow*, many felt the train should be built. Most said they didn't have the money to do it because of the need for new equipment as World War II came to a close. Besides, building materials were scarce, and some people felt that no one would be able to get permission from the government to build the train. However, by the summer of 1945, the war seemed to be nearing an end, and government officials were encouraging private industry to begin developing innovative new projects that would show America's strength and stability. It was "arranged that the Executive Committee come out to Chicago to view the model train. With the reactions to the Burlington car as a background and the fact that so many railroad executives had expressed the wish that we build such a train, [General Motors chairman of the board Alfred P.] Sloan and [GM president C. E.] Wilson felt that we should proceed with the construction of these cars. During the discussion on the advisability of undertaking this large investment, naturally many points of view were expressed, particularly because General Motors had no intention at that time nor does it have now, of entering the car manufacturing business."[1] In the fall of 1945, GM's Executive Committee decided to build the *Train Of Tomorrow*. "The reaction of these railroad leaders has been highly favorable. In view of their general agreement that the *Astra Liner* represents a long forward step in the advancement of railroad travel enjoyment and comfort, we have reached the decision to build four demonstration cars. It is felt that a further contribution to rail transportation can be made by carrying through the complete engineering and construction of such a train, in conformance to the ideas developed by the original designers. [The *Astra Liner*] will be built by General Motors as quickly as war material and manpower conditions permit."[2]

After "several of the established builders were invited to submit proposals on construction of the *Train Of Tomorrow*, the proposal of the Pullman-Standard Car Manufacturing Company was accepted,"[3] and a contract was signed with Pullman-Standard on November 6, 1945, to build the four cars according to specifications and standards set down by the American Association of Railroads (AAR). The big surprise is that GM did not actually submit any drawings, blueprints, or specifications for the construction of the cars to Pullman-Standard. According to the specification book completed by Pullman-Standard in July 1946, "Although no construction specification had been furnished, if deemed necessary, General Motors reserved the right to assist in the design." However, it was the interior design that was supplied by GM. "On these cars, the design of the interior finish is as dictated by the

Laying the foundation, center sill, and crossbeams of *Star Dust*. At this point the undercarriage is upside down. Pullman Technology, Inc.

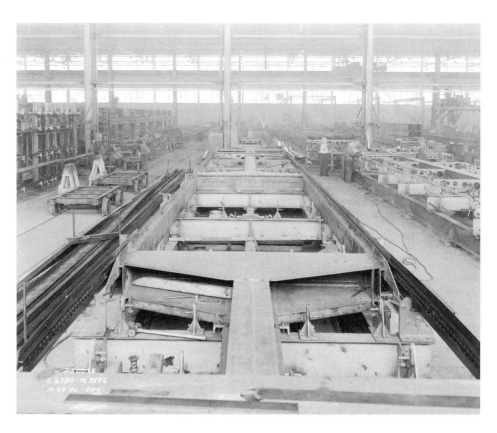

The attachment of the sidewalls of the chair car, *Star Dust*. Pullman Technology, Inc.

General Motors Styling Section, and particular attention was paid to producing the effects desired, as shown on color drawings prepared by the General Motors Styling Section."[4]

Inspectors supplied by EMD were given quarters at the plant and, according to the specification book, "were given access to [Pullman-Standard's] drawing room, calculations and etc., and any department where any material for these cars might be manufactured and the inspectors were furnished with complete up-to-date prints of drawings and specifications and revisions when made. Any material or workmanship rejected by the Electro-Motive inspectors was corrected or replaced."

EMD engineer Martin Blomberg acted as liaison with Pullman-Standard, and drawings and specifications as well as any changes made in the design and construction of the cars were sent to him for his approval.

Before actual construction was begun, full-scale wooden mockups were constructed to test measurements and make adjustments in the designs. Then a final set of drawings and specifications was produced by Pullman-Standard. Arrangements were made with the War Production Board to get the steel necessary for construction, but the windows in the dome would still have to be made from flat glass because all curved glass production was still needed for the war effort. "All features entering into the design or construction of these cars on which Pullman held patents were supplied without any royalty charge."[5] General Motors did not charge any royalty fees to any other car builders or railroads for the use of the dome idea or any other ideas used on the train that didn't directly involve GM products, such air-conditioning, refrigeration, roller bearings, or power-generating units. Construction started in October 1946 and was completed May 17, 1947.

Pullman-Standard published a brochure called *Train Of Tomorrow: Built by Pullman-Standard*, which gave the following details about the construction of the cars:

> The *Train Of Tomorrow* cars are built primarily of low-alloy high-tensile steel, using Pullman-Standard welded-girder design with certain modifications in the underframe and roof framing to accommodate the Astra Dome on each car.

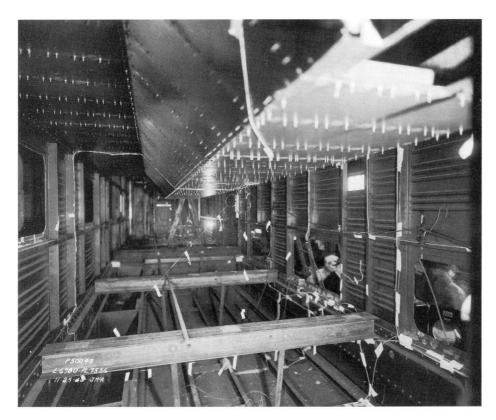

Construction of the floor of the dome section of *Sky View*. Pullman Technology, Inc.

The dome area under construction on *Sky View*, facing the front of the car where the waiter's station will be built. Pullman Technology, Inc.

Star Dust on a Pullman-Standard plant rolling platform. Pullman Technology, Inc.

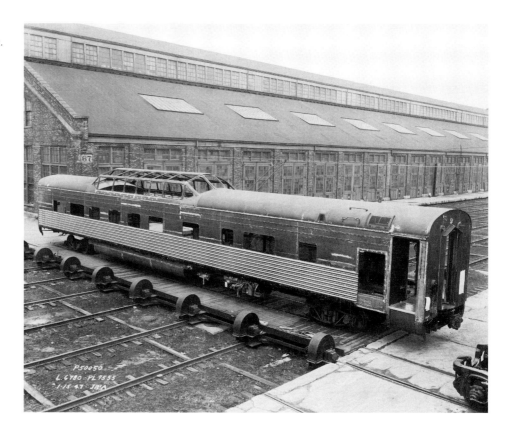

The exterior dome of *Star Dust*.
Pullman Technology, Inc.

To provide for the lower-level floor under the dome, which is depressed into the underframe, it was necessary to cut off the longitudinal underframe members at each end of the dome section where they are attached to heavy built-up arc-welded box-section beams, placed crosswise of the car. These beams are connected at the side post lines to built-up welded box-section members, placed longitudinally, which transmit buffing loads through this section of the car. The box-section members line up horizontally with the center sills to avoid eccentric loading, and their cross-sectional area exceeds that of the center sills. Zee and angle side sills extend the full length of the car and are riveted in the dome portion to the box-section members.

The floor of the lower level is supported on built-up arc-welded H-beams, placed crosswise of the car and extending up and connected to the side sills. Pressed Z-shape longitudinal stringers are placed between the beams. Bottom sheets of steel are arc-welded to the underframe members so as to tie together all underneath framing at this portion of the car.

In the roof, the rolled Z-shape side plate is replaced with a built-up arc-welded box section extending the full length of the car. This reinforced sideplate section carries the compression stresses in the roof and makes up for the loss of material in the roof because of the cutout for the dome.

The dome roof members between the windows are box sections made up of flanged channels spot-welded together except at the car center line where the air duct is located. All roof members are fitted and arc-welded together, and the whole unit is arc-welded to the main roof members. These members do not carry any stresses from static loading in the car, flexible connections to the main roof members having been provided at certain points to accomplish this result. The dome roof construction is equal in strength to that of the balance of the car roof.

The upper floor in the dome is supported by a structural member extending the full length of the dome and the full width of the car. This member is made up of double sheet steel with spacers between placed crosswise of the car; spacers and sheets are spot-welded and riveted together.

The completed assembly is securely fastened to the side framing of the car so that in addition to supporting the floor, it acts as a tie between the two sides for the length of the dome where the regular roof framing has been cut away.

To prove the strength of the design, several physical tests were conducted using as a specimen the car structure for the sleeping car. These tests include a static load test, torsion jacking test, and AAR compression or squeeze test. The first two were conducted in the Pullman-Standard plant under the joint direction of engineering department representatives of the Electro-Motive Division and Pullman-Standard. The third test was conducted at the Altoona, Pennsylvania, test plant under the supervision of AAR representatives.

In the first test, a 100 percent overload in the car failed to show excessive stress on any of the 250 strain gauges applied at critical points throughout the structure. Most gauges showed stresses lower than expected.

The left side of *Sky View* on shop trucks at the Pullman-Standard plant. Pullman Technology, Inc.

The kitchen area of *Sky View* looking toward the rear of the car and the dome area. Pullman Technology, Inc.

In the second test, performed to check the stresses and deflection when jacking a fully loaded car at the four corners with the added condition of having one jack fail, no gauge showed excessive stress and there was little deflection at the unsupported corner of the car.

In the third test, the car showed an average deflection between body bolsters of 0.485 in. when a maximum load of 839,396 lbs. was applied on the center-sill ends, this deflection being well within the limits set up in the AAR specifications.

The squeeze test, load test, and jacking test, also called the torsion test, were conducted on the sleeping car *Dream Cloud*. The car was taken to Altoona, where it passed the squeeze or crush test, involving applying a load in excess of 800,000 pounds, the AAR requirement. On static load tests, a 200 percent overload was applied with no damage to the car.

When the cars were completed, they were taken on a test ride from Chicago to Wallace Junction, Indiana, and back on the Monon Railroad. According to the specification book, "In addition to the above tests, such tests as required for the different systems in-

A stress test is being performed on *Dream Cloud*. Pullman Technology, Inc.

stalled in the cars, such as water, heat, air-conditioning, lighting, etc., were conducted and [Pullman-Standard] made the required changes to insure the systems perform as intended."

Not many of the details about the construction of the locomotive 765 are known. The Electro-Motive Division of General Motors built the E7A diesel locomotive, but it is not known at which of EMD's three plants construction took place. The locomotive was built as part of order 765, hence the number used on the locomotive on tour. Other than exterior changes, such as the addition of corrugated stainless steel on the lower body, three stainless steel moldings at the letterboard area near the top of the side body to match the moldings on the cars, and *Train Of Tomorrow* emblems of the front and side of the body, the locomotive was just a standard diesel locomotive. The reason for this was so that the public and any railroad executives inspecting the unit would see what a locomotive in regular service would be like.

E-units were passenger locomotives with two diesel power-generating units and three axles on each truck, instead of two as on freight units. E7A passenger locomotives had two

A jack test is being performed on *Sky View*. Pullman Technology, Inc.

1,000-horsepower 2-cycle, 12-cylinder v-type GM diesel engines. Power to the trucks was transmitted by an electric transmission that consists of a direct current generator driven directly by each engine to the traction motors in each set of trucks that are geared directly to the axles. The locomotive was 71 feet long and had a loaded weight of 318,000 pounds.

The unfinished locomotive was exhibited at an open house at the Fisher Ternstedt plant in Cleveland, Ohio, on April 25, 1947, just one month before it would begin pulling the *Train Of Tomorrow* on its 28-month tour.

E. Preston Calvert, a public relations consultant to Pullman-Standard at the time the train was built, recalled that "Pullman-Standard built the shell for the locomotive. At the time [GM] told Pullman-Standard they would like them to build all the shells for the locomotives, but Pullman said it did not want to get into locomotive building."[6]

Although General Motors never published the price it paid for the train, it was widely believed at the time to be between $1 and $1.5 million. Some people at GM today have said that only Alfred P. Sloan, C. E. Wilson, and a few accountants may have ever known how much was spent for the train. Even then, there may have been hidden costs in the budgets of various departments and divisions.

CHAPTER 4

SOMETHING FOR EVERYONE TO SEE

Having just returned from a preview run to French Lick, Indiana, the *Train Of Tomorrow* had been cleaned and moved to a track near Soldier Field for the christening ceremonies. The Grand Ballroom of the Palmer House Hotel was all set up for a luncheon for 1,000 invited guests. The breasts of chicken, potatoes, and peas were in the kitchen waiting to be served. A cake in the shape of the *Train Of Tomorrow* stood in front of the rostrum where such luminaries as Alfred P. Sloan, C. F. Kettering, Cyrus R. Osborn, and Chicago mayor Martin H. Kennelly would speak.

That was the stage setting on May 28, 1947, when the *Train Of Tomorrow* was to be inaugurated before starting its nationwide tour.

Actually, the whole thing started two days earlier when the train left Dearborn Station on the Monon route of the Chicago, Indianapolis and Louisville Railroad for a preview run to French Lick and back with some of the nation's leading newspaper and magazine writers, radio personalities, and about 75 General Motors, Pullman-Standard, and Pullman Company executives. Having left Chicago at 10:30 AM on May 26, the train arrived at the private track of the French Lick Springs Resort 30 minutes late, at about 5:15 PM, due to some courtesy stops along the way. There was only one minor incident on the first leg of the trip, near Greencastle, when the train ran over some tie plates on the track.

A dinner was held for all who had made the trip, and then the guests stayed overnight at the famous spa. The train was opened for inspection by the guests of the hotel from 8:00 to 10:30 PM. During that period, 758 people toured the train; other than the trip itself, this was the first public showing. The next morning the train left French Lick at 10:10 AM, 10 minutes late due to some lost luggage. Scheduled to arrive back in Chicago at 4:30 PM, the train didn't pull into the Illinois Central station until 6:15 PM.

Besides being a preview for the press and executives, the trip was seen as a shakedown run to work out all the "bugs" before the train opened for its public exhibition and tour. For example, on the return trip from French Lick, several brake tests were made from high speed. "New passenger cars are always checked more closely for brake performance than anything else."[1]

On the trip between Chicago and French Lick, the streets of the towns and cities were

The *Train Of Tomorrow* arriving at French Lick, Indiana, on May 26, 1947, during the preview run. On the left side of the photograph, notice the ramps for the locomotive lying on the grass. GM.

Dinner is served at the French Lick Springs Resort for the dignitaries on the preview run. They spent the night at the resort, returning to Chicago the following day. GM.

Felix Dorian Sr. broadcasts live over WGN Chicago from the observation lounge car, *Moon Glow*, on the preview run. GM.

lined with people wanting to be among the first to get a glimpse of the train. Lunch was served onboard, and passengers in groups of two were invited to visit the locomotive.

The return trip was filled with unusual events. During a stop for water in Bloomington, the framework of the domes was struck by a chain hanging down from a trackside water tank. A stop was made for the photographers to make a profile shot of the train. At 2:30 PM, the train struck an automobile at a crossing. There were no injuries, and the locomotive had very little damage, but the hood of the car was knocked off and its side was crushed. Then, approaching Dyer, Indiana, Cyrus Osborn made a 30-minute call through the train's radiotelephone system to the RMS *Queen Elizabeth*, 2,400 miles away at sea. Everyone on the train heard the call, as it was broadcast over the public address system. The steamship's captain, C. G. Illingworth, wished the train's passengers luck with the new adventure, and Osborn described the new train to the steamship's passengers.[2]

Upon arrival back in Chicago, the train was sent to the Illinois Central coach yards where it was cleaned and sent through a carwash before it was moved to a track near Soldier Field and made ready by 2:00 PM on May 28 for inspection by GM's special guests.

Around noon, 1,000 leaders in business, finance, and politics as well as executives from General Motors and its various divisions, Pullman-Standard, and the Pullman Company sat down to a luncheon sponsored by the Electro-Motive Division. "One assistant manager of the Palmer House said, 'the luncheon was one of the greatest turnouts in the history of Chicago.'"[3] Lew Diamond and his orchestra entertained the guests while they ate, then the speeches began, officiated over by Cyrus Osborn.[4] Following the close of the ceremonies at the hotel at 2:30 PM, the guests were loaded onto buses and taken to Soldier Field for the christening ceremonies. Jane Kettering, granddaughter of C. F. Kettering, waved a heat device that started the locomotive's engines and smashed a bottle of champagne against the step at the engineer's door on the locomotive.[5] Guests were then invited to tour the train.

From May 29 through June 1, 50,000 people stood in line in the rain at Soldier Field, waiting to see the *Train Of Tomorrow*. On June 2, the train was cleaned and serviced,

The *Train Of Tomorrow* hit this Chevrolet at a crossing between Bloomington and Dyer, Indiana. GM.

This is the couple whose car was struck by the *Train Of Tomorrow* on the preview run. She looks like she is thinking, "Sue the bums!" There is no record as to what happened as a result of the accident. GM.

The damage done to the front skirt of the locomotive as a result of the collision with the Chevrolet. GM.

Jane Kettering, granddaughter of Charles Kettering, prepares to christen the *Train Of Tomorrow* by breaking a bottle of champagne on the step to the engine room of locomotive 765. GM.

News and Sound Master film crews prepare to capture the action of the inauguration ceremonies of the *Train Of Tomorrow* at Soldier Field in Chicago. GM.

switched to the Grand Trunk and Western, and moved to Track 18 at the Dearborn Station annex to prepare for departure at 9:15 AM for the start of what was believed would be a 6-month tour. At that time no one could predict that the tour would actually last 28 months, traveling over 65,000 miles.

There were lots of details involved with moving the *Train Of Tomorrow* on its tour. Employees and volunteers from General Motors and its Electro-Motive, Frigidaire, Delco, Hyatt Bearings, GMAC, and AC divisions, plus those from Pullman-Standard, the Pullman Company, and the railroads over which the train ran and was exhibited, worked behind the scenes, preparing the advertising and publicity, handling the accounting, and servicing, maintaining, cleaning, and operating the train. People in every city volunteered to help. Vendors made deliveries of food and supplies to the train along the tour, while hotels supplied accommodations for the train's traveling staff. The number of people involved and the details to be handled might have seemed impossible to manage at first, but through a lot of hard work and dedication to the dream, it all came together in what many consider one of the great events in railroad history. Other trains have been on tour, before and since, but nothing was like the thrill of the *Train Of Tomorrow*. It was as if people could not get enough.

The first step in conducting the tour was taken by the advance man, who went to a city or town ahead of schedule to make all the necessary arrangements for the train's arrival and exhibition. Public relations expert Eddie A. Braken recalls that promoting the train "was like advancing a show of Barnum and Bailey. You do what you can to build it up."[6] Preston Calvert adds, "It was easy to promote anything when a company says, 'Here's a million bucks. Go promote this train, and when that's gone, there is more behind it.'"[7]

The advance man would contact all the local GM people and enlist their help. In fact, it was said that the people from General Motors Acceptance Corporation were responsi-

NEWS OF THE DAY

On May 28, 1947, the christening of the *Train Of Tomorrow* was not the only news story of the day. In fact, it was a very small part of the world's events. The Chicago newspapers reported that the Senate had voted to cut taxes and the Trailways bus drivers had gone on strike. The cost of living in France was up 1,000 percent while wages were up only 500 percent. In addition, the citizens of Newfoundland were hoping to align themselves with the United States instead of being sold by the British government to Canada.

The weather forecast for the Chicago area was cloudy with showers and thundershowers during the evening hours. The high temperature was 58, with a low of 50. In fact, a heavy fog settled over the Chicago area during the night, and many airplanes had to be talked down through the fog.

On the editorial page of the *Chicago Tribune*, opinions were expressed about the freedom of the press and support of a seaplane base near the Navy Pier.

In sports, the Pittsburgh Pirates had defeated the Chicago Cubs 7 to 6. It was also reported that 28 cars had qualified for the Indianapolis 500 to be run the following Sunday.

The business news of the day was that late season rain and snow were hurting farmers in the corn belt and that a judge had cleared the way for the Gulf, Mobile and Ohio Railroad to take over the Alton Railroad on July 1.

The radio schedule listed *The Great Gildersleeve* as being on WMAQ at 7:30 PM. At 8:30, *The Dinah Shore Show* played on WBBM with Peggy Lee and Peter Lind Hayes as guests. And at 9:00, the *Philco Radio Time*, featuring Bing Crosby, had child star Margaret O'Brien as a guest.

The only television station listed, WENR, showed *Frank Wood, Private Detective* at 8:00 PM and then signed off at 10:00 after being on the air for only four hours.

The comics showed the detective Dick Tracy dashing off to pick up a doctor to deliver a baby at Mr. Lacy's house. In *Brenda Starr*, a puppy named Tornado was acclimating himself to the "good life" at Mrs. Palmer's apartment, only to turn up his nose at a small bone, after dreaming of having a bone bigger than himself.

At the movies, Claudette Colbert and Fred MacMurray were starring in *The Egg and I*. The Apollo Theater was showing *The Jolson Story*, while the Oriental had the film *Dillinger* (banned until now!), as well as a stage show featuring Leon Navara and his orchestra, the acrobatic Maxellos; the trio Joe, Lou, and Marilyn Cates, and the Allan Sisters.

The classified ads listed an eight-room brick house in Hyde Park, near the University of Chicago, for sale at $16,500. The ad said the house had enclosed porches, sanded floors, steam heat, and a garage. In the employment classified ads, Old Line Legal Reserve Company was looking for two insurance agents with a income potential of $90 per week in commissions. For the ladies, Illinois Bell Telephone Company had a large ad requesting applications for operators at $40 per week for a five-day workweek.

Display ads showed Pepsi available at many places for 6 cents a glass, and a complete filet mignon dinner was available at Tracy's on the Avenue for $1.90. Two film stars were shown in quarter-page ads endorsing cigarettes: Joan Crawford for Raleigh and Barbara Stanwyck for Chesterfields. Gentlemen could purchase white broadcloth shirts at Maurice L. Rothschild's for $4.50, and American Airlines was selling seats on their 65-minute flight to Detroit aboard a DC-6 for $13.60, plus tax.

In spite of the fact that the story about the inauguration of the *Train Of Tomorrow* did not appear until the next day, General Motors, Frigidaire, and Pullman-Standard ran ads about the public display of the train at Soldier Field on May 28. The ads announced that the train would be on display from 2:00 until 9:00 PM May 29 through June 1.

---

ble for much of the work done on the tour. Almost every area of the country has a local GMAC representative who helps with making financial arrangements for customers. However, General Motors sees them as their representatives at the local level, which can be not only a conduit of information back to the company but also a resource to be used for help with the corporation's projects locally. At many of the small towns and communities through which the train passed without stopping, the local GMAC representative helped build interest in the train, to the point of getting local people to be trackside when the train passed through. Many schoolchildren made field trips just to see the *Train Of Tomorrow* slow down and pass through their hometown. People on-board would throw pamphlets to the crowd. "You would get the educators to take the children down to see the train," says Braken. "When the train came to that little town, and there were 700–800 kids from the schools out there, the train would slow down while it went through and everybody would wave. That was nice. That was good promotion."

When the train was visiting a city that had a GM plant, the train might be exhibited there. Frequently the plants held an open house for the employees and their families and then would act as hosts for the general public to take a tour of the plant facilities as well as the train. When the train was in a plant city, but was on exhibit at a location other than at

The *Train Of Tomorrow* traveling through Lawrence, Kansas, on a run from Topeka to Kansas City on April 5, 1948. Wallace W. Abbey Collection—All rights reserved.

the plant, the morning hours would sometimes be reserved for the GM employees and their families to take a tour. Employees from the plants and other local GM offices, dealerships, and facilities would serve as hosts while the train was opened for public tours. School groups were also invited to tour the train during the morning hours. It was reported that quite a few children coerced their parents into taking them back for another look.

The advance man would also build interest among the local press, and a special press run would be arranged. The reporters would attend a preview luncheon to hear about the train and the reasons for the tour, and then they would be bused out to a point 30–50 miles from the city and ride the train back. They were given tours of the train as well as press releases and photographs to be used in publicizing the train's visit to their city.

The train on exhibit at the Wazee Market in Denver on November 5, 1948. GM.

Arrangements would also be made for a place to exhibit the train, and local volunteers would be enlisted to help with the exhibit. The police would be notified, and arrangements would be made for special police coverage to help with security and crowd control. Retired GM executive Dick Terrell remembers, "Detroit had a lot of fancy ideas about where they wanted it displayed. The marketing people and the styling people thought that it ought to be in some kind of real scenic setting with big mountains in the background and snowcapped, and my reaction was, 'Well, how in the hell do you do that in a railroad station in Pittsburgh?'"[8]

Occasionally, state and local politicians would be invited to participate in the festivities and the press runs. They would be photographed in the cab of the locomotive, sitting in the engineer's seat. This was good publicity for them and appeared as an endorsement of the train.

When not hosting tours, the morning hours were usually for cleaning and maintenance. Frequently local VIPs and GM executives were invited to have lunch in the dining car on the days the train was in a city for exhibition. Public tours began about 1:00 or 2:00 PM and were scheduled to end at 9:00 PM. Sometimes the line would be so long at closing time that the crew would keep the train open until the last person had been through the train. Sometimes this meant the train crew would not be able to finish until after midnight.

At times the GM employees would be required to perform public relations duties by speaking at preview luncheons or by making presentations to civic organizations.

Visitors standing in line in *Sky View*. Almost 6 million people walked through the train while it was on tour for General Motors. GM.

Standing in line to walk through the train on a visit to Detroit. Ironically, the site is across the street from Ford's Highland Park plant. GM.

General Motors employees, along with their families and friends, stand in line at the Detroit Transmission Plant's open house on September 27, 1948, to take a tour of the train. GM.

The *Train Of Tomorrow* being readied for exhibition. GM.

THE *BLUE GOOSE*

The *Blue Goose* was used as a dormitory car for the Pullman crew, a locker area for the General Motors employees on tour, and a baggage car for storage of the ramps, supplies, and literature needed when the four-car *Train Of Tomorrow* was on exhibit.

Blue Goose was originally called *National Road*, commemorating the first national toll road, which was built in the early nineteenth century from Cumberland, Maryland, to central Illinois. According to William Kratville's *Passenger Car Catalog*, the remodeled Monitor-type heavyweight car was built in 1925 from Plan 3951, Lot 4885 by Pullman-Standard as one of 18 club-buffet-baggage cars used on Baltimore and Ohio Railroad's *National Limited*.

The car usually moved ahead of the *Train Of Tomorrow*. From time to time, though, inconvenient railroad schedules or limited service to an exhibition city caused the *Blue Goose* to be added to the consist just behind the locomotive. Panels painted silver below the belt line and three silver stripes near the roof line were added to match the exteriors of the four dome cars.

According to Dick Terrell, "They used to carry the *Blue Goose* on other passenger trains. I never worried about that. It was such a simple problem. We would break the show up and take the train to another city. So I would ask the railroad just to put the *Blue Goose* on a train and get it there, and that's what they would do. That was one of our easier problems."

The sitting and locker area for the Pullman and General Motors crews on tour. Notice the two brass spittoons under the shelves near the door. EMD—GM.

The left side of the *Blue Goose* from the dorm area at the vestibule end of the car. EMD—GM.

Looking from the sitting room to the area for the berths. EMD—GM.

The right side of the *Blue Goose* from the forward end of the car. EMD—GM.

The nickname *Blue Goose* had been given to the car by the *Train Of Tomorrow* crew, but it was not used officially until the legal right to use the name had been granted. The source or reason for the granting of the legal permission to use the name is unknown. A nameplate using shadow painting to match the lettering on the other cars was then added to the car in December 1948 or January 1949.

So little is known about the history of the car, and not much is known about its movements during the tour. The car is mentioned five times in the *Train Of Tomorrow*'s daily log (see appendix A). As a matter of fact, the car was not mentioned in any schedule until the second Canadian tour in the fall of 1949, just before the train ended its tour.

From an EMD memo about the Canadian tour's schedule, dated September 9, 1949, from G. W. Rukgaber (the first train manager) to C. R. Osborn, is the schedule of the movements of the *Blue Goose* during the train's second tour of eastern Canada, from September 20 to October 30, 1949:

Tuesday, September 20, 1949–
Wednesday, September 21, 1949
Leave Chicago 11:00 PM CT GT 6–28
Arrive Detroit 7:50 PM ET
Baggage-dormitory car to be moved on the first convenient train and by ferry to the CNR Windsor Station.

THE *BLUE GOOSE* continued

Wednesday, September 21, 1949–
Thursday, September 22, 1949
Leave Windsor   11:05 PM ET CN 10
Arrive London   1:50 AM ET

Sunday, September 25, 1949
Leave London   5:55 AM ET CP 20
Arrive Toronto   8:50 AM ET
Leave Toronto   4:00 PM ET Pool 6
Arrive Ottawa   10:00 PM ET Pool 560

Thursday, September 29, 1949
Leave Ottawa   7:30 AM ET CN 48
Arrive Montreal   10:45 AM ET

Tuesday, October 4, 1949–
Wednesday, October 5, 1949
Leave Montreal   11:00 PM ET Pool 358
Arrive Quebec   5:25 AM ET

Sunday, October 9, 1949
It will be satisfactory to the General Motors people to handle the baggage-dormitory car immediately behind the diesel from Quebec to Sherbrooke.

Tuesday, October 11, 1949
It will be satisfactory to the General Motors people to handle the baggage-dormitory car immediately behind the diesel from Sherbrooke to Montreal and placed on Pool 5 from Montreal into Oshawa.

Friday, October 14, 1949
Leave Oshawa   6:16 AM ET Pool 17 or Pool 19
Arrive Toronto   7:15 AM ET

Thursday, October 20, 1949
Leave Toronto   7:10 am ET CN 77
Arrive Hamilton   8:30 am ET

Sunday, October 23, 1949
Leave Hamilton   9:15 am ET CN 102
Arrive St. Catharines   10:08 am ET
In view of the limited service from St. Catharines to Stratford, General Motors is agreeable to handling the car on the *Train Of Tomorrow* for this portion of the journey.

Thursday, October 27, 1949
Leave Stratford   3:00 am ET CN 39
Arrive London   3:50 am ET
Leave London   4:00 am ET CN 9
Arrive Chatham   5:45 am ET

Friday, October 28, 1949
Leave Chatham   5:58 am ET CN 9
Arrive Windsor   7:15 AM ET

After 11:00 PM ET on October 29, the entire *Train Of Tomorrow*, including the baggage-dormitory car, should be delivered to the Michigan Central Railroad for movement to Detroit, Milwaukee Jct., and Chicago.

Sunday, October 30, 1949
Leave Detroit   2:30 AM ET GT
Arrive Chicago   8:30 AM ET

At the end of the tour, the car was sold to the Rock Island Railroad, and it became Locomotive Instruction Car 1820.

The GM employees on tour with the train would stay in hotels, and the Pullman crew would sleep in a baggage-dormitory car that accompanied the train. Sometimes the GM crew would sleep on the train when it was moving from one city to the next on an overnight run. The Pullman crew responsible for a particular car would also sleep on-board during this type of movement.

When the train arrived at an exhibition site, a set of specially made ramps would be set up. The public entered the train through the rear door in the observation lounge of the *Moon Glow* and exited through the vestibule of the chair car, *Star Dust*. Ramps were also set up so the public could visit the cab of the locomotive. The ramps could be set up on either side of the train depending on the exhibition site. Besides the four dome cars and the locomotive, there was a baggage-dormitory car, sometimes referred to as the office car and nicknamed the *Blue Goose*, which traveled either ahead of or with the train. The car's baggage compartment was used for storage, and the seating area on the car had been converted into a dormitory for the Pullman crew and for storing the GM staff's personal belongings. When the exhibition ended, the ramps were dismantled and placed on the *Blue Goose* for transport to the next exhibition site. In addition to ramp storage, the baggage room on the office car was used for the storage of pamphlets and brochures as well as extra parts.

As Dick Terrell stated in a 1988 interview, "The railroads provided all the operating crews: engineers, firemen, brakemen, flagmen, and conductors. They were all railroad

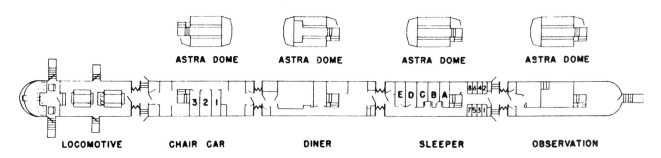

This is the layout of the *Train Of Tomorrow* on exhibit. Notice the placement of the stairs. Guests entered through the rear of the observation lounge car, *Moon Glow*, and exited through the front of the coach car, *Star Dust*, before going on to see the locomotive. GM.

The *Train Of Tomorrow* arriving at Lawrence, Kansas, at 11:25 AM on April 4, 1948. Wallace W. Abbey Collection—All rights reserved.

The train on exhibit at the Santa Fe station at Lawrence, Kansas, on April 4, 1948. Over 7,000 people toured the train that afternoon. This is a typical setup for public exhibitions. Wallace W. Abbey Collection—All rights reserved.

people. Pullman's responsibility was to provide the dining car staff and car attendants. As far as the maintenance of the train, the supervising of it was all a General Motors function, and that was handled by the EMD. They paid for it; it was theirs and they had the responsibility."

Besides supplying car porters, waiters, chefs, and inspectors, the Pullman Company was responsible for the cleaning and some of the maintenance while the train was on tour. In a May 1947 memorandum to agents and district superintendents, the Pullman Company defined the responsibilities of Pullman to GM:

> General Motors has asked our forces to take care of the inside and outside cleaning of the four cars, plus painting and upholstering attention and inside mechanical, electrical, air-conditioning, and heating features. Our men will also make whatever repairs and do what servicing is necessary to the lighting, air-conditioning, and heating features under the cars, as requested from time to time by General Motors representatives accompanying the train. We will also take care of the same features on the GM office car, except where the exterior cleaning is done by the railroad where this car operates on regular trains. In addition to the above, we will furnish help for setting up and taking down ramps that will be used during the exhibition. From time to time the exhibit train may be put through the railroad's exterior car washer or arrangements made by the GM representative to have the exterior cleaned by the railroad forces, in which event we will not do the cleaning; this is to be determined by GM representatives accompanying the train.
>
> The railroad forces will do the regular maintenance of trucks, air brake, and draft gear on the cars and do all of the locomotive work including cleaning the interior and exterior. This arrangement is tentative and will be subject to minor changes from time to time as requested by GM representatives.

From time to time a special crew car for cleaners and maintenance men traveled with the *Blue Goose*, which was used strictly for the Pullman and GM train crews.

Even though the *Train Of Tomorrow* was invited by all the railroads it traveled on, government regulations forced General Motors to file a special tariff that had to cover the entire tour and pay $2.25 per mile for the train and the office car. Plus, they had to pay coach fares for all the passengers who were not entitled to free transportation. On the runs when

This advertisement by Pullman-Standard appeared in several newspapers and in color in many of the national magazines at the start of the *Train Of Tomorrow*'s tour. The advertisement is the property of Pullman Technologies and is reprinted with written permission. Pullman Technology, Inc.

the crew slept on the train, step-up tickets had to be purchased to cover the cost of the accommodations. Dick Terrell recalled, "Tickets and step-up tickets were needed for the *Train Of Tomorrow*. In the case of our people, who were there to service the equipment, we had some kind of operating agreement with the railroad, and I think we had passes for those people. But if you just started to load people on there, I would assume that we had to pay some fares, and that would be very legitimate and legal; nothing wrong with that."

Some of the highlights of the tour include:

- In one 150-hour period, 150,000 people toured the train.
- After seeing the *Train Of Tomorrow* at the Railway Supply Manufacturers Association exhibition in Atlantic City, June 23–27, 1947, George Peck, a reporter for the *Paw Paw* (Ill.) *Times*, wrote, "Let me tell you that the railroads are here to stay."

SOMETHING FOR EVERYONE TO SEE

- One small town in Pennsylvania closed all of its shops and stores so that everyone could watch the train pass through.[9]
- C. W. Carey, from Silver Spring, Maryland, an 84-year-old locomotive engineer who had retired after 50 years with the Baltimore and Ohio Railroad, "took a long, lingering look at the train [and] admitted wistfully, 'Boy, I was born 50 years too soon.'"[10]
- Life on-board the train for the staff could be very trying at times, especially at the beginning of the first tour. In an 1988 interview, Eddie A. Braken remembered that he "didn't get home for a change of clothing for three months. I was buying clothes on the road." His hard work paid off when "during the stay in D.C., I was promoted to regional manager. As a matter of fact, I was promoted in an Astra Dome. Paul Garrett, then the vice president for PR, had me, when we were parked there in Silver Spring, come up into one of the domes, where we could have a little privacy. We talked and he told me I was being promoted to regional manager and sent to Buffalo. Of course, I wasn't long in Buffalo before the bloody train was coming to town through Syracuse and Rochester."
- Fresh anthuriums, a rather expensive but exotic flower that helped set the mood of the train, were aboard all the time. Preston Calvert recalled, "There was only one place we could get them. That was the Palmer House Flower Shop. They would fly them in and place them throughout the train."
- When something is famous or important, the power of getting things done is exhibited, as when the train was backing into Johnstown, Pennsylvania, on October 24, 1947. While backing through a tunnel into the railroad yard of a steel plant, the last piece of curving track supposedly shifted a fraction of an inch, causing one of the chrome strips on the dining car *Sky View* to be rolled back just a bit. During the night, workmen moved the track so the train could leave through the same tunnel the next morning without causing any more damage to the train.
- The *Train Of Tomorrow* made two trips to southeastern Canada during the 28-month tour. Between September 1 and September 8, 1947, it was put on exhibit at the Canadian National Exposition. Then it visited the McKinnon Industries in St. Catharines. The second Canadian Tour was 1949 near the end of the train's tour. The train crossed the border at Windsor on September 22, visiting London, Ottawa, Montreal, Quebec, Sherbrooke, Oshawa, Toronto, Hamilton, St. Catharines, Stratford, Chatham, and Windsor before leaving Canada on October 30.
- The *Train Of Tomorrow* was exhibited during a three-day open house at the EMD plant in LaGrange, Illinois, from October 24 to October 26, 1947. However, it was taken off display for one day to do a special run for C. E. Wilson, president of General Motors. The train left LaGrange late on October 24 and proceeded to Detroit, arriving early the next morning. The train was cleaned and then boarded by Wilson and the editors from the Associated Press for a trip to Ann Arbor for a football game between the University of Michigan and the University of Minnesota. Following the game, the guests were serenaded by the University of Michigan marching band. The train arrived back in Detroit at about 6:00 PM, leaving for LaGrange at 8:45 PM and arriving at 4:45 AM, just in time for the open house on Sunday.
- While on tour in Colorado on the western tour, the train made a special run with 25 models to Colorado Springs for a day of publicity photography in the mountains, including Pikes Peak.
- A "celebrity run" was made on December 16, 1947, from Glendale to Saugus and returning. Besides a number of GM executives and press agents, the train was host to such Hollywood celebrities as Ginger Rogers, Walter Pidgeon, Art Linkletter, Eddie Bracken, Joan Leslie, Herbert Marshall, "Wild Bill" Elliott, Jean Hersholt, and Ann Sothern. The food was catered by Giro's.
- The University of Michigan played the University of Southern California in the Rose Bowl on January 1, 1948. The *Train Of Tomorrow* played a part in the whole

The *Train Of Tomorrow* on exhibit at the EMD open house in October 1947. Note the other EMD equipment on display in the background. EMD-GM.

The *Train Of Tomorrow* at the EMD open house. EMD—GM.

The *Train Of Tomorrow* leaving Ann Arbor, Michigan, on October 25, 1947, following the Michigan-Minnesota game with C. E. Wilson, president of General Motors, and the editors of the Associated Press on-board. GM.

festivities surrounding that game. On December 28, the Michigan football squad was taken for a ride from Glendale and Saugus and back before the train left for San Francisco to pick up the university's 149-piece marching band, returning to Glendale on December 29.

Then on January 1, the train made a run to Santa Barbara to pick up the USC football team. Train staff members were guests of USC for the game.

Art Linkletter and Ginger Rogers in the dome of the *Sky View* during a celebrity run on December 16, 1947. GM.

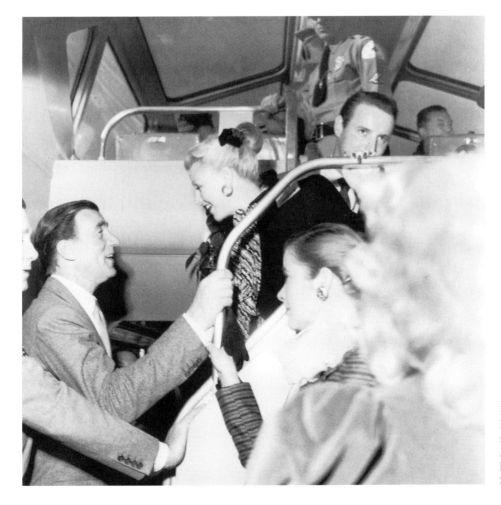

Herbert Marshall, Walter Pidgeon, and Ginger Rogers meet on the stairs to the dome of *Moon Glow*. The blonde with her back to the camera is Ann Sothern. GM.

Danish radio and film actor Jean Hersholt and his wife enjoy the view from one of the big windows in room 3 located under the dome of the chair car, *Star Dust*. GM.

Actor Herbert Marshall in the dining car *Sky View* having a drink with his bride of four months, Patricia "Boots" Mallory. GM.

Eddie Bracken, "Wild Bill" Elliott, and Joan Leslie in the dome of *Moon Glow*. GM.

The *Train Of Tomorrow* in California's Feather River Canyon during the western tour. GM.

47

- At Wickenburg, Arizona, on January 7, the train's manager and engineer were abducted in an old-fashioned train robbery by masked bandits as part of that town's celebration of the train's visit.
- On a run from Phoenix to Albuquerque on January 11, Santa Fe Road foreman Osburn pushed the locomotive to the limit. From milepost 328 to 315 east of Flagstaff, the train was clocked doing 138 miles per hour (26 seconds per mile), even though the speedometer registered only 118½ miles per hour.
- In Albuquerque on January 12, Jack Benny and party visited the train while the eastbound California Limited was on a 15-minute stop. Then on January 26, the German shepherd actor Rin Tin Tin visited the train while it was being exhibited in Austin, Texas. He posed for a few pictures and then left his "stamp of approval" on the skirt of the locomotive.
- Another train on tour around the United States at the same time was the original *Freedom Train*. The two trains met in Waco, Texas, on February 2, 1948. The *Train Of Tomorrow*'s manager, George Rukgaber, visited the *Freedom Train* and then invited his counterpart, Walter H. S. O'Brien, to take a tour of the *Train Of Tomorrow*. Departure from Waco was delayed 20 minutes while O'Brien completed his tour.
- The *Train Of Tomorrow* arrived in Tulsa, Oklahoma, on March 24, 1948, and was greeted with a ceremony by the Otoe Indians, who named the train "Usah-Paw-Pitsi-Su," meaning "swift pony with a glass saddle." A public relations man on the tour from General Motors, Johnny Johnson, was given a real war bonnet and anointed "Petah Turah-hi," meaning "chief who sees far."[11]
- The train visited many plants and other GM facilities but made only one special trip to an automobile dealership. While on exhibit in Flint, Michigan, the train

An aerial shot of the *Train Of Tomorrow*'s visit to L. C. Anderson's new Chevrolet dealership in Lake Orion, Michigan, on July 6, 1948. This was the only dealership that the train visited. GM.

The *Train Of Tomorrow* and the Baltimore and Ohio's antique locomotive meet at the 1948 Chicago Railroad Fair. EMD—GM.

made a trip to Lake Orion, just north of Detroit, on July 16, 1948, to a track across the street from Lee Anderson's Chevrolet showroom. There was a parade followed by an open house of the train at the dealership's facilities.

- The *Train Of Tomorrow* was on display at both the 1948 and 1949 Railroad Fairs in Chicago. The first fair was from July 20 to September 24, 1948, on the lakefront in Chicago. The 1949 fair was from June 25 until October 2, 1949.
- Editors from all over the country were invited to make the first run of the train in 1949. On January 17 the train made a run from Detroit to New York City for the opening of the GM Auto Show at the Waldorf=Astoria. Going for a speed record, the train left Detroit at 7:30 AM, making the trip across southern Ontario doing speeds up to 100 miles per hour. Short stops were made at St. Thomas, Ontario, Buffalo, East Buffalo, Elmira, and Scranton. Through New York State and the hills of Pennsylvania, the train reached 102 miles per hour. With a party of about 110 people, three meals were served, and an accordionist supplied entertainment. The train reached Hoboken at 7:30 PM, making the trip in a flat 12 hours taking 114 minutes off the best previous time. During the switching operations for a run to Rocky Mount, North Carolina, at 1:00 AM, passengers in the Astra Domes could spot the Statue of Liberty in the moonlight.
- Two unrelated but nevertheless interesting events took place on May 31, 1949. On the press run to Springfield, Illinois, a Mr. Stubbs, a publisher, suffered a heart attack but recovered enough to be taken to his home. Following the train's arrival at the Illinois Central freight yards in Springfield, Cyrus Osborn was presented with a plaque printed with a joint resolution from the Illinois legislature. The Sixty-sixth General Assembly of the Illinois legislature passed joint resolution 32, congratulating GM and Pullman for designing and building the *Train Of Tomorrow* in Illinois.

THE CHICAGO RAILROAD FAIR

The Chicago Railroad Fair, held in observance of the centennial of the birth of railroad service in Chicago, was held on a 50-acre parcel of Burnham Park on Lake Shore Drive between 20th and 30th streets near Soldier Field. (The site had been used for the 1939 Century of Progress exhibition and is now occupied by the McCormick Place convention center.) The first fair was held from July 20 through September 24, 1948, and was repeated by popular demand from June 25 through October 2, 1949. Admission was 25 cents, and more than 4 million people attended the fair during the two summers.

A lapel pin from the 1948 Chicago Railroad Fair. Ric Morgan Collection.

Cover of the September 1948 issue of *Railroad Magazine*. Carstens Publications, Inc.—All rights reserved.

Railroad service in Chicago began October 21, 1848, when the Chicago, Burlington & Quincy started runs on 12 miles of branch line track north from Aurora to Turner Junction with the *Pioneer* locomotive borrowed from the Galena and Chicago Union Railroad.

Thirty-eight railroads as well as car builders and suppliers participated in the fair. There were such exhibits and attractions as an "Old Faithful" geyser replica and rodeo at the combined display of the Northern Pacific, Great Northern, and Burlington Railroads; Illinois Central's re-creation of the French Quarter in New Orleans; a San Francisco cable car ride; Santa Fe Railroad's Indian village; an ice show; the Cyprus Gardens Water Thrill Show; the Chicago and Eastern Illinois "Florida in Chicago" exhibit with a southern mansion and tropical fruit trees; the Pullman display; a puppet theater; a nightly fireworks display; and the Gold Gulch Opera House, where two railroad-related melodramas were performed. There was a narrow gauge line called the Deadwood and Central City Railroad that ran three-quarters of a mile from the main entrance to the south end of the fairgrounds and back. At the EMD display, an F3 diesel freight A unit was featured with the body sides, doors, and machinery covers removed, allowing the public to see the inner workings of a modern locomotive. In 1949, a 35-foot animated Paul Bunyan was added that moved its lips as well as its arms, head, and eyes in synchronization with recordings. Besides the *Train Of Tomorrow*, which was displayed both years, the "Famous Trains" display featured Baltimore and Ohio's *Train X* in 1948 and a locomotive and a set of cars from the Spanish government's new "Talgo" mystery train in 1949.

During the 1948 fair, places to eat were limited but included refreshment stands for hot dogs and hamburgers and a couple of restaurants for full-service meals. A few railroads had dining cars on display that were open to the public for lunch and dinner, serving full-course meals. After several complaints, the food service facilities and menus improved for the 1949 fair.

Of all the attractions, the highlight was the "Wheels a-Rolling" pageant, staged in a specially built amphitheater seating 5,000 people on the lakefront and expanded in 1949 to seat 6,000. The pageant illustrated the role played by the railroads in the development of the United States. With Lake Michigan as a backdrop, the performances had a cast of 150 people dressed in full costume as Indians, trappers, Yankee peddlers, itinerant preachers, the population of a frontier garrison, soldiers, cowboys, early settlers, Harvey girls, railroad presidents, Civil War soldiers, dandies, road gangs, and women of the Gay Nineties re-creating events in transportation in the Midwest from the seventeenth century until modern times. Two announcers narrated the story, split up into segments or scenes. The high point was the use of actual railroad equipment onstage ranging from the historic *Rocket* brought to the United States in 1827, to the modern diesel locomotive, including the *Best Friend of Charleston*, the *Tom Thumb*, the *Lion*, the *Atlantic*, the *Lafayette*, and the *William Mason* as well as the locomotive that started it all for Chicago as a railroad town, the *Pioneer*. Eighteen locomotives or trains were used as well as 20 early automo-

THE CHICAGO RAILROAD FAIR continued

biles, 36 oxen and horses, stagecoaches, covered wagons, and antique railroad passenger cars. Three tracks were laid on the 450-foot by 100-foot stage with a mile and a half of track behind the scenes to maneuver the trains on and off the stage.

Produced from a script written by railroad historian Edward Hungerford, the production was designed by Helen Tieken Geraghty, who had directed the transportation pageant at the 1939 Century of Progress. The technical director, Arthur Mayberry, had worked in the same capacity with the pageants at the Cleveland and New York World's Fairs. There were four shows a day: 2:00, 4:00, 7:15, and 9:00 PM. Admission was 60 cents for the one-hour show. New scenes as well as more vehicles of historical significance were added in the 1949 edition of the pageant.

The *Train Of Tomorrow* was on display at both fairs in Chicago. The full-time crew took vacations on an alternating schedule but remained with the train while it was exhibited. Thirty-six college boys, hired as hosts for the train, were rotated in their positions so they wouldn't get bored with their jobs. The Burns Agency guarded the train at all times.

The one problem mentioned in the train's log was the lack of good eating establishments, a problem corrected the next year. During the fair in 1948, sandwiches and coffee were served from the train's diner to the staff and hosts hired for the duration of the fair.

Only once during the 1948 fair did the train leave Chicago. On September 8 the train made a special run to Dayton to take the press for a day at Indian Lake. There were 224 people on-board, and the lounge car was packed beyond capacity. The train returned to the fair the next day.

For the 1949 fair, facilities were improved. Pavement had been extended to 12 feet from the tracks to the outside stanchions. The stationary stanchions were equipped with nine floodlights. The site for the second year was close to the grandstands and backstage area of the fair's pageant.

State of Illinois
Sixty-sixth General Assembly
Senate
Senate Joint Resolution No 32
Offered by Senator Bidwill

Whereas, General Motors' Train of Tomorrow, featuring scores of engineering advancements and refinements, including the famed Astra Domes or glass "penthouses" atop the cars, will arrive in Springfield on May 31 and will be on public exhibition in this city on June 1 and 2; and

Whereas, The Train of Tomorrow may truly be called an Illinois product, as the idea of the Astra Dome was conceived by an Illinois resident, Mr. C. R. Osborn, vice-president of General Motors and general manager of its Electro-Motive Division at La Grange, Illinois, the original model of the train was first exhibited in Chicago, and the train itself was constructed in this State by Pullman-Standard Car Manufacturing Company; and

Whereas, Although not in the railroad car building business, General Motors, because of its interest in the improvement of all forms of transportation, designed the train and ordered the cars built as a contribution to increased comfort and safety in rail transportation; and

Whereas, The Train of Tomorrow is a symbol of the free enterprise system and represents the American capacity for getting things done by methods that guarantee the individual freedom that makes a people strong; therefore be it

A copy of the joint resolution presented by the Illinois legislature to C. R. Osborn on May 31, 1949, to honor the *Train Of Tomorrow*'s visit to Springfield. EMD—GM.

51

> *Resolved, By the Senate of the Sixty-sixth General Assembly of the State of Illinois, the House of Representatives concurring herein, that, on behalf of the People of the State of Illinois, we cordially welcome the Train of Tomorrow to Springfield, the capitol of Illinois, and that we commend and congratulate General Motors and all other companies and individuals who participated in the construction and development of the Train of Tomorrow.*
>
> *Adopted by the Senate, May 5, 1949*
>
> *[signed]*
> President of the Senate
>
> *[signed] Edward H. Alexander*
> Secretary of the Senate
>
> *Concurred in by the House of Representatives, May 12, 1949*
>
> *[signed] Paul Powell*
> Speaker, House of Representatives
>
> *[signed] Chas. F. Kerwin*
> Clerk, House of Representatives

The bill was introduced by State Senator Arthur Bidwill and adopted by the Senate on May 5 and by the House of Representatives on May 12.

- Romance bloomed on the *Train Of Tomorrow*. George Rukgaber, the first train manager, married Doris Shotwell, a GM secretary from New York City, after meeting her at the Albany, New York, stop of the tour in September 1947. They were married during the 1947 Christmas tour break.
- Ed Fish, who worked for Arthur Kudner, the institutional advertiser for GM, wrote a jingle called "The Wonderful *Train Of Tomorrow*." A record was made and sent around to the radio stations where the train would make tour appearances. Eddie Braken recalled, "The disc jockeys loved it. It wasn't a great song, but it was at the right time. It was used to help publicize the train."[12]
- The last trip of the train on October 30, 1949, was made from Windsor, Ontario, through the tunnel to Detroit, and then on to Dearborn Station in Chicago, where the train had begun its preview run 28 months earlier. From there the train was taken to the 103rd Street EMD plant, where it was refurbished and placed in temporary storage. The final entry in the daily log ends: "Our last ride on the Blue Lady. The *T.O.T.* has been home to most of us for two and one-half years. Looking back on these many months of close association with other staff members makes

SOMETHING FOR EVERYONE TO SEE

The music and lyrics to "Wonderful *Train Of Tomorrow.*" Ed Fish.

one pause and silently salute each and every one. They were gentlemen, all.... So long, Blue Lady, for us you will always be the *Train Of Tomorrow.*"

When the tour ended on October 30, 1949, the *Train Of Tomorrow* had traveled over 65,000 miles, visiting 181 towns and cities. While it is estimated that 20,000,000 had seen the train, 5,761,307 either walked through or rode it. There was never any mention of what it cost to take the train on tour; some sources believe it may have run anywhere between $1 and $2 million, and that does not include the cost of the locomotive and four cars. The *Train Of Tomorrow* had led an exhilarating life on tour. Now General Motors had to decide what to do with it.

CHAPTER 5

FOR SALE: ONE TRAIN

THE *TRAIN OF TOMORROW* WAS FOR SALE FROM THE TIME IT WAS BUILT. ORIGInally, the train was expected to be on tour for about six months and then be sold to a Class 1 railroad.[1] The feeling was that the cars should go together. It was not essential that the four cars be placed on the same train, but it was felt that this would be best.

One statement in a packet of information called the "*Train Of Tomorrow* disposal program" said, "It is felt that the project [has] accomplished its purpose. It is, therefore, recommended that the train program be terminated following the close of the [Chicago] Railroad Fair."[2] However, the recommendation was not followed, as the train went on tour in Canada before being refurbished and placed in storage at EMD's 103rd Street plant.

On October 17, 1949, Cyrus Osborn received an interoffice letter about the idea of donating the amount in excess of the asked selling price to a national charity. The sale was planned this way: All Class 1 railroads that were interested in purchasing the train would submit bids based on an asking price of $950,000 for the locomotive and four dome cars, plus all equipment and spare parts. The highest bidder would take possession of the train. In the event two or more bidders offered the asking price or higher, the one with the highest bid over the asking price would get the train and the amount over the asking price would go to charity. When the idea was bandied about, everyone thought the idea was a novel one and the railroad would certainly get some great publicity. But there were problems. Someone pointed out that, according to Interstate Commerce Commission accounting practices, the excess amount could not be charged as an operating expense. Also, there was a feeling that the directors of some of the railroads or holding companies could not vote for the use of stockholders' money for charitable purposes without the consent of a majority of the stockholders and that would be a very difficult task. However, there was some indication that this would not be a problem at all railroads, and if it was seen that there was a problem, the company wouldn't bid or the obstacle would have been cleared before a bid was placed.

On October 21, Osborn sent a three-page letter offering the *Train Of Tomorrow* for sale to the presidents of 29 railroads.[3]

As with anything like this, there were some problems with the sale. In itself, the train did not have enough cars to make it profitable to run on its own, and most of the railroads stated they didn't have a run they could put it on exclusively. If the train was used on overnight service, an identical train would be needed for operation in the opposite direction, and the *Train Of Tomorrow* was considered a unique piece of equipment. It would

GENERAL MOTORS CORPORATION

LA GRANGE, ILLINOIS

VICE PRESIDENT'S OFFICE

October 21, 1949

Within the next several weeks, the General Motors TRAIN OF TOMORROW will have completed its tour of public exhibition throughout the United States and Canada. Since the Train was first placed on public exhibition in May, 1947, it will have visited 181 cities, travelled 65,000 miles, have been seen by 20,000,000 people, more than 5,761,307 of whom have either walked through or travelled in the Train. As a part of its tour, it was on exhibition at the Chicago Railroad Fair during the summer months of 1948 and 1949. The interest and favorable reaction of the public towards the Train has far exceeded our original expectations. Therefore, in building and exhibiting the Train, we feel our original purpose of stimulating public interest in railroad travel has been amply fulfilled.

It is now our intention to offer the complete Train to the railroads for sale. As indicated in the attached descriptive data, the Train consist is as follows:

1 - E-7 2,000 h.p., E.M.D. Diesel Locomotive

1 - 72-Passenger Lounge Car

1 - 52-Passenger Seating Capacity All-Electric
      Dining Car

1 - 20-Bed Sleeping Car
      48 Seating Capacity

1 - 68-Passenger Observation Car

The Train, because of its use for exhibition purposes, has been kept in first-class condition throughout its entire tour. Originally built by Pullman, it has been returned to the Pullman organization approximately every six months for such refurbishing and other work required to maintain it in absolutely new condition.

ELECTRO-MOTIVE DIVISION

C. R. Osborn sent this letter to the presidents of 29 railroads on October 21, 1949, offering to sell the *Train Of Tomorrow* equipment. Notice the asking price of $950,000. UPRR.

Page 2   October 21, 1949

The Train is completely equipped for immediate operation, including specially designed silverware, china, blankets, bar equipment, kitchen utensils, etc. It is intended that all such accessories and equipment shall be included in the price and sold with the Train.

A supply of repair parts on all special equipment, such as the Diesel engine generator sets, air conditioning, electric kitchen equipment, and other items of a similar nature, will be available through the Electro-Motive Division.

As of November 1st, the Train will be parked at the Electro-Motive Division Plant No. 2, 103rd Street and Cottage Grove Avenue, South Chicago, and will be open for inspection to all railroad prospective purchasers.

We have had requests from some twenty railroads that we offer them the Train for sale, many of these requests being on the basis of first refusal.

In order that we handle the matter as equitably as possible for all railroads, we have decided to offer the complete Train, including the locomotive, for bid, with an upset price of $950,000. By this we mean, we will entertain any bid from any railroad but should one or more railroads bid more than $950,000, we will accept from the high bidder the price of $950,000 and the amount in excess of that sale price will be donated by the successful bidder himself during the year 1950 to any single national charity which it may select. Should the unusual situation arise of two or more railroads with the highest bids offering the same exact sum, we shall try to work out with the railroads involved some equitable and agreeable means to meet this situation. We should like to have these bids in our hands by December 1, 1949.

The amount specified above does not, of course, approximate the original cost of the Train to us nor its replacement value today. There is also the added factor of the great publicity this Train has received through national advertising, newspapers, radio, and the people who have seen and visited the Train. It is reasonable to expect, therefore, that the eventual owner will be able to capitalize from this publicity through the revenue producing capacity of the Train. For this reason, it is agreeable to us, if the final purchaser so desires, to continue to operate the Train with all of its original emblems and markings of the General Motors TRAIN OF TOMORROW. We will extend the right to any railroad to use the name "TRAIN OF TOMORROW" as long as they may wish and, coupled with this, the right to the use of the General Motors name and emblem for a period of five years, which we feel should be ample time to capitalize the publicity value which exists for the Train.

We are attaching clearance diagrams of the Train and layouts of the various cars. We are also prepared to furnish construction drawings of the cars if they should be desired.

After the bids are received and the eventual sale consummated, the Train will be available for immediate delivery.

Very truly yours,

C. R. Osborn

Omaha – March 18, 1950

521-3

Memorandum by President:

Referring to attached AFE request for the purchase of General Motors' "Train of Tomorrow".

General Motors has expressed willingness to sell the train intact to the Union Pacific for $500,000, which is considerably less than it could obtain for the train by the sale of individual components for which it has obtained numerours bids, and as we can use the train to good advantage to improve our passenger service between Portland and Seattle, about which we have had numerous complaints, I consider it an advantageous purchase. This will, of course, necessitate some modification of the present Portland-Seattle pool train arrangement, but I am confident this can be equitably accomplished.

The attached AFE includes an amount of $100,000 to cover expenses in connection with the purchase and possible contingencies, including freight.

While the train has been run some 60,000 miles in exhibition, it has been maintained in excellent condition and can be dilivered to us promptly at Omaha.

(Signed) A. E. Stoddard

A. E. Stoddard sent this memorandum to the Union Pacific Executive Committee in New York on March 18, 1950, encouraging them to purchase the *Train Of Tomorrow* for use in the Northwest corridor between Seattle and Portland. UPRR.

also take a long time to build the second train, and most railroads did not want to wait very long once the purchase had been approved. Another problem was that many railroads simply didn't have the money. Most of them had made huge capital investments after World War II to purchase new equipment to update their rundown fleets. In addition, passenger traffic was beginning to drop. Quite a few people had discovered the convenience and speed of air travel. The airlines were also increasing the size of their fleets to handle the demand, and new airports were being built to accommodate the increase in air passenger traffic.

While the *Train Of Tomorrow* was never considered a "white elephant," it was probably more difficult to sell than was originally expected. No records were found stating how many bids had been placed, but Osborn's letter had stated that all bids should be in the hands of EMD by December 1, 1949. However, it wasn't until the spring of 1950 that the train was sold.

In a memorandum dated March 18, 1950, Union Pacific president A. E. Stoddard recommended approval of the purchase of the *Train Of Tomorrow* to UP's Executive Committee. The memo stated that Union Pacific could purchase the whole train, including the locomotive, for $500,000 and that an additional $100,000 should be approved for freight charges, refurbishing, and any other miscellaneous costs related to the purchase of

FOR SALE

FORM A. F. E. 12

UNION PACIFIC RAILROAD COMPANY
AUTHORITY FOR EXPENDITURE

Request No. 35

Executive Committee, New York.
Authority is requested for the following Expenditure for Investment in Road and Equipment.

Submitted March 18, 1950 (Signed) A. E. Stoddard
President

CHARACTER OF WORK AND REASONS WHY RECOMMENDED

LOCATION: EQUIPMENT

DESCRIPTION: Purchasing General Motors Corporation "Train of Tomorrow" consisting of -

1 - 2000 HP Diesel-electric locomotive.
1 - 72 passenger Astra-Dome lounge car.
1 - 52 passenger Astra-Dome all-electric dining car.
1 - Sleeping car having 3 compartments, 2 drawing rooms, 8 roomettes and 24 seats for passengers in the Astra-Dome.
1 - 68 passenger Astra-Dome observation car.

REASON FOR
RECOMMENDING: To augment present fleet of streamline trains and to effect improvements in passenger service between Portland and Seattle.

BUDGET REFERENCE: Not in budget.

CHARACTER OF
IMPROVEMENT: No. 40½ - Streamlined - Passenger Trains

1. Total estimated expenditures - - - - - - - - - - - - - - - - -	$ 600 000	
2. LESS—(a) Amount chargeable to Operating Expenses - - - $		
3. (b) Value of Salvage chargeable to "Material on Hand" $		
4. Estimated cost of additional property, chargeable to Investment in Road and Equipment for which Executive Authority is required - - - - - - -	600 000	
5. Appropriated for expenditure during calendar year 19...... - - - - - - -	—	$ 600 000

Approved by the Executive Committee at New York, N. Y.

MAR 21 1950   19___

(Signed) E. G. SMITH
Secretary.

On March 21, 1950, the Executive Committee of the Union Pacific Railroad elected to purchase the *Train Of Tomorrow* equipment. This document is the authority to make that purchase and provides extra money for freight and the reconditioning of the train. UPRR.

the train. Stoddard based his recommendation on the need to upgrade the equipment for passenger service between Portland and Seattle, where the railroad had received a number of complaints. The memo stated that EMD had received numerous bids but was willing to sell the train intact for $500,000. What that probably means is that either all the bids were lower or the other bidders were after individual pieces of the equipment and EMD decided not to sell cars individually. On March 21, the UP Executive Committee approved the purchase of the *Train Of Tomorrow*.

CHAPTER 6

GOING TO WORK

O<small>N APRIL 6, 1950, THE UNION PACIFIC RAILROAD ISSUED A "WORK ORDER AU-</small>thority," which is a listing of estimated expenditures, for the *Train Of Tomorrow*. The $600,000 total allocated for the purchase of the train was allocated as follows:

Locomotive	$ 57,282.91
Astra Dome chair car	99,996.67
Astra Dome all-electric dining car	114,993.30
Astra Dome sleeping car with three compartments, two drawing rooms, and eight roomettes	114,206.84
Astra Dome observation car	113,520.28
	$500,000.00
Freight	2,000.00
Power equipment	10,200.00
Repairing train; removal of carpet; alterations in electrical controls on power units; train electrical equipment and electro-pneumatic brake; and miscellaneous expenses, including contingencies	87,800.00
	$600,000.00

After the train was delivered in April, Union Pacific painted its exterior in the yellow, red, and gray of the railroad's livery, replaced the carpet, and refurbished the whole interior. A large taillight was added to the roof of the observation car *Moon Glow*. In May 1950 the train went on a tour of the UP system, concentrating on the Northwest, where the cars would be placed in pool service with cars from the Northern Pacific and Great Northern between Seattle and Portland. There was a lot of speculation in the press as to what the Union Pacific would do with the individual cars or the complete train. However, right from the start, officials with the railroad had decided to use the train in Seattle–Portland service. Following the tour, the train served as a second section on the *City of Los Angeles*.

At 8:00 AM on June 18, the four cars, with other streamliner cars of Train 457, headed

The *Train Of Tomorrow* in Union Pacific livery. UPRR.

The former *Train Of Tomorrow* on display at the Omaha Union Pacific station during a tour of the railroad's system in May 1950. UPRR.

The new Portland–Seattle train on display at the Tacoma Union Station in 1950. UPRR.

The new Portland–Seattle train on display at the Portland Union Station in 1950. UPRR.

north over the Pacific Coast Route of the Northern Pacific Railroad to Seattle, arriving at 1:05 PM. After nearly four hours of layover, during which a passenger could conduct business or explore, Train 458 departed Seattle at 5:00 PM, arriving back in Portland at 9:30 PM. The consist of the train in both directions was usually three E7 locomotives, a mail-baggage car, five or six 44-passenger day coaches, a regular parlor car, a through sleeper from the *City of Portland*, and the four dome cars from the *Train Of Tomorrow*. One jour-

Northbound Train No. 457 approaching the Tacoma–Narrows Bridge area. Notice the large red light added to the back of the observation car, *Moon Glow*, later renumbered UP 9010. UPRR.

Northbound Train No. 457 running along the Columbia River at Kalama, Washington. Note the private car just behind the baggage car. The dome cars are near the rear of the train. UPRR.

nalist remarked, "The tourist trade gets a magnificent daily view of Mt. Hood, the Columbia River, the Olympic and Cascade Mountains and Puget Sound, and a thrilling streamliner ride as well."[1] By August 13, the schedule had been shortened so that the train was able to make the 182-mile trip in 3 hours and 59 minutes. Adjustments in times and schedules were made over the years.

The round-trip coach fare for the trip in 1950 was $6.84, tax included. For a seat in the dome of *Star Dust*, the fare was $9.49. To get a reserved seat in the parlor car or a reserved day seat in *Dream Cloud,* the sleeper, was $15.65, which included the use of *Moon Glow*. The dome diner *Sky View* was open to all passengers. One magazine article at the time noted that the parlor car fare was $2.64 cheaper than a round-trip flight on a DC-6 between the two cities.[2]

In March 1956, all the cars were renumbered except for *Dream Cloud*, which continued in service with its name intact. *Star Dust* was renumbered 7010, and the dining car *Sky View* became 8010. However, being renumbered 9015 was not all that had happened to the *Moon Glow*. In December 1958, the car was reclassified from a dome observation car to a dome lounge car. In 1959, the rounded rear end was cut off, and the end was squared so the car could be used in midtrain service. As a lounge for parlor car passengers, *Moon Glow* was not getting a lot of use. Railroads have frequently put a lounge car next to the diner, as a place for dining car customers to wait, and that was the purpose of making the change on *Moon Glow*.

The cars would not see service for many more years. Occasionally the cars were used on special moves, not only on the UP but on neighboring railroads as well. Documents show the *Moon Glow*, the lounge car, being used on a special called *Advance 104*, from Los Angeles to Chicago, via Omaha.[3]

Southbound Train No. 458 arriving at Portland's Union Station. UPRR.

Sleeper *Dream Cloud* awaiting dismantlement in November 1966. Harry Stegmaier Collection—All rights reserved.

Having been removed from regular service on March 20, 1962, *Dream Cloud* was put back into service on the Portland–Seattle trains during the World's Fair in Seattle. The car last saw service on October 1, 1962.[4]

Dining Car 8010, formerly the *Sky View*, was the first to be retired in February 1961 with a depreciated salvage value of only $225.[5] *Dream Cloud* was retired in February 1964 with an appraised salvage value of $2,624.[6] The chair car *Star Dust* was valued at $3,390 when it was vacated in November 1964.[7] *Moon Glow* was the last to be vacated in March 1965, with an appraised value of $2,254.[8] While the records for each car were marked "vacant but not dismantled," all the cars were eventually sold for scrap to McCarty's Scrap Yard in Pocatello, Idaho.

As for the fate of the locomotive from the *Train Of Tomorrow*, it was renumbered as unit 988 and placed in general service, but not necessarily used on any train that carried the four dome cars. It was supposedly sent back to EMD in 1965 and rebuilt as a more powerful E9A unit and then renumbered 912. There are some who believe that the E7A unit was actually traded in on a new E9A unit rather than being rebuilt, but the records at EMD and UP do not state which is the case.

The story of the *Train Of Tomorrow* does not end here . . .

CHAPTER 7

LIFE AFTER DEATH

THE FATE OF THE *Star Dust, Dream Cloud,* AND *Sky View* WAS THE SCRAP YARD blowtorch, but the *Moon Glow* has survived to lead still another life after serving on the *Train Of Tomorrow* and seeing passenger service on the Union Pacific Railroad.

Since Pocatello was the maintenance yard of the Union Pacific for passenger equipment in the West, the four cars had probably been put into retirement there. Union Pacific sold all of the cars to a scrap yard in Pocatello, Idaho, owned by Bill McCarty.[1] In 1969 Henry Fernandez bought an old livestock auction lot next door to McCarty's Scrap Yard, converted it to a scrap yard, and named it Henry's Scrap Metals. Fernandez saw one of the cars from the *Train Of Tomorrow,* the *Moon Glow,* at McCarty's and thought he would like to have it because he "could see it was unique," and it fascinated him "because it was so far advanced," even though he didn't know its history or that it had been a part of the *Train Of Tomorrow*.[2] After purchasing the dome car for $2,500, Fernandez had the Dale Wood House Movers relocate it to his scrap yard, where he planned to have the car restored and converted into an office. However, it wound up sitting there for 18 years before it was rescued.

Riding south on 30 West in Pocatello while on vacation in June 1978, David Seidel saw the car in the scrap yard as he drove by. He turned his car around and went back to Henry's, but the gate was locked. Deciding to spend the night in Pocatello, he went back to Henry's the next day, getting a chance to see the car and take some photographs, one of which appeared in a national rail fan magazine a few years later. Upon returning home, he did a little research and discovered the car was *Moon Glow* from the *Train Of Tomorrow,* after which he sent Fernandez a letter with the information and a copy of the photographs. This was the first time in almost 10 years Fernandez knew anything about the history of the car.

Murl Rawlins Jr. saw the car for the first time in 1971. While visiting family in Pocatello, Rawlins saw the car every year, but he never realized its significance until June 1980, when he and several members of the National Railway Historical Society (NRHS) made a fan trip to the UP yard and shops in Pocatello. The group included Lester Tippe, president of the Promontory Chapter, Dan Kuhn, George Swallow, Rawlins, and his father.. The group left Salt Lake City on June 20 with most of the members returning home late Saturday. However, a few chose to stay another night in Pocatello. On Sunday morning Murl Rawlins Sr. said, "Let's show Dan the dome car in the scrap yard."[3]

They arrived at Henry's scrap yard about 9:00 AM. Even though the yard was closed,

The front end of *Moon Glow* (UP 9010) at Henry's Scrap Metals in Pocatello, Idaho, in June 1978. David Seidel Collection—All rights reserved.

they could still see the car through the fence. Kuhn knew instantly that it was one of the cars from the *Train Of Tomorrow*. "I stood there for several minutes pondering how such a famous and historically significant piece of American transportation history could have gone unnoticed for so long in a scrap yard in Pocatello, Idaho. My first thought was, 'This car must be preserved!' It was several months before I learned which of the *Train Of Tomorrow* domes I was looking at. That Sunday it didn't matter. All that was important was that one *did* exist, and I knew we had to get it out of that scrap yard and restored."[4]

Upon their return to Salt Lake City, the club members began discussing the situation with the Railroad Advisory Board of Ogden Union Station Museum (now the Utah State Railroad Museum) and the museum director, Teddy Griffith. They asked Fernandez to donate the car to the club, but he refused. He did invite them to see the car, though. So, in June 1981, Tippe, Kuhn, Rawlins Jr., and Swallow inspected the car and again discussed its sale or donation. By that time Fernandez knew what the car was and placed a high value on it. Fernandez admits now that he didn't know what he wanted to do: keep the car or make the donation. However, he did promise not to destroy the car and to contact the Promontory Chapter in the event he decided to get rid of it. The group visited Fernandez's accountant to make another proposal, but nothing came out of all these meetings for about three years.

Due to some problems with the Internal Revenue Service, Fernandez decided to close the scrap yard. So he called Lester Tippe one day and "said that he had decided to dispose of the car and would we please put together an appraisal as to the value of the car, which we did, based on historical value, cost of restoration, and the original cost."[5] Fernandez and his wife, Ardy, signed over the car to the Promontory Chapter on December 27, 1984.[6] On February 1, 1985, Rex L. Firth, vice president and general manager of the Salt Lake, Garfield and Western Railway Company, sent a letter to Tippe establishing a value of $138,000 on the car, which is the amount of the donation allowed to Fernandez.

As Fernandez made plans to close down his scrap yard, an auction was planned to sell off all the contents of the compound. He informed the Promontory Chapter that they had to get the car out of the scrap yard right away, or else it might be sold in the auction. The

HENRY'S

P.O. BOX 291 • POCATELLO, IDAHO 83201 • 3633 GARRETT WAY • 208 233-4315

December 27, 1984

The National Railway Historical
Society, Inc., Promontory Chapter
Salt Lake City, Utah

KNOW ALL MEN BY THESE PRESENTS:

That we, HENRY FERNANDEZ and ARDY FERNANDEZ, by these presents do donate and convey to the NATIONAL RAILWAY HISTORICAL SOCIETY, INC., PROMONTORY CHAPTER,

One (1) Railroad Domeliner Passenger Car, Observation Lounge known as the "Moonglow" car built by Standard-Pullman Company in Lot #6780, the last example of the four cars built in 1946. This car originally toured the country in the "Train of Tommorrow",

And, we will warrant unto the National Railway Historical Society, Inc., Promontory Chapter, that said railroad domeliner car is free and clear of any lawful claims and demands of all and every person whatsoever.

Witness our hands and seal this 27th day of December 1984.

_____
Henry Fernandez

_____
Ardy Fernandez

_____
Notary Public

_____
Address

My Commission Expires Jan. 10, 1988

_____
Witness

_____
Accepted by
For The National Railway Historical
Society, Inc.   Promontory Chapter
Salt Lake City, Utah

This is the letter transferring ownership of *Moon Glow* (UP 9010) from Henry Fernandez to the Promontory Chapter (Salt Lake City) of the National Railway Historical Society. Henry Fernandez Collection—All rights reserved.

members of the Promontory Chapter then began to look for a place to put the car, but they had little success until they discovered the Ogden Union Station Museum was willing to take the car and restore it, but the museum would have to be responsible for moving the car from Pocatello to Ogden. The people at the museum agreed if they could get ownership of the car. An agreement was reached, title was transferred, and plans to move the car got under way.

The first thing was to hire someone to move it out of Henry's scrap yard. The man who had moved the car the first time, Dale Wood, was contracted to move the car for about $3,000, as well as supply a set of trucks at additional cost. In February 1987, he moved the car to a siding behind E. J. Bartell's, an insulation company located at milepost 333 on Highway 30 West in Pocatello about a quarter of a mile north of Henry's Scrap Metals,

Moon Glow before it was moved out of the scrap yard by the Dale Wood House Moving Company. Henry Fernandez Collection—All rights reserved.

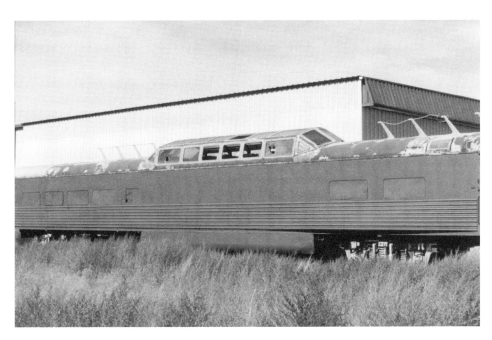

Moon Glow at the siding of the E. J. Bartell Company in Pocatello on September 9, 1987. Ric Morgan Collection.

The Lower Lounge looking toward the Upper Lounge. The car was repeatedly vandalized after it was removed from the scrap yard. Ric Morgan Collection.

The bar in the Lower Lounge facing the rear of the car. Ric Morgan Collection.

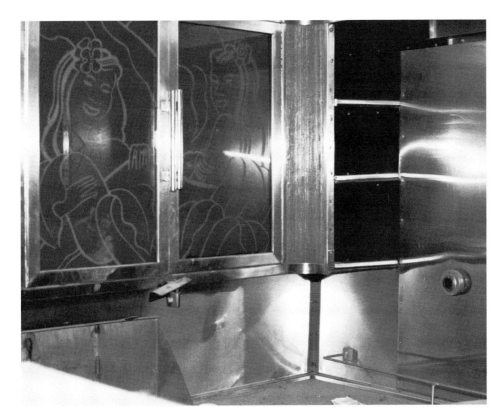

Some of the original etched glass on the bar survived through the years. Ric Morgan Collection.

The lower bar storage areas still look good after nearly 17 years in the scrap yard. Ric Morgan Collection.

The dome seating area of *Moon Glow*. The turquoise leather on the dashboard is intact, as is most of the glass. Ric Morgan Collection.

The writing desk and electronics storage cabinet located in the observation lounge has changed little since the car was built. The clock is missing, as are the electronics behind the doors above the desk. Ric Morgan Collection.

The observation lounge looking toward the front of the car. The lighted table in front of the built-in sofa has been rebuilt since the car was moved to Ogden. Ric Morgan Collection.

The observation lounge looking toward the rear of the car. Notice that the rounded ceiling was maintained when the observation end was cut off by the Union Pacific Railroad in 1956 so that the car could be used in midtrain service. Ric Morgan Collection.

OGDEN UNION STATION

Moon Glow has found a home at the Ogden Union Station. The station was given to the city in 1977 and now houses the Utah State Railroad Museum, Myra Powell Gallery art museum, the Wattis-Dumke HO scale model train museum, the Eccles Railroad Center, the Browning Firearms Museum, the Browning-Kimball Car Museum, a new natural history museum, and a restaurant, as well as being the Amtrak station for Ogden. The building is also used as a cultural center for the city by providing the M. S. Browning Theater, the Old Timers Hall, and conference rooms. The building is equipped with a kitchen that can serve up to 1,000 diners.

Built in the Mediterranean style, Ogden Union Station was formerly owned by the Ogden Union Depot and Railroad Company, a holding company for the Union Pacific and the Southern Pacific and built in 1924 on the foundation of the original station, which had burned down a few years earlier.

Ogden Union Station received a grant of $150,000 from the Bicentennial Commission and $600,000 from the State of Utah in 1976 to refurbish the building and help set up the facilities for public use. Renovation was completed in 1979 and is now registered with the National Registry of Historic Places.

The railroad museum equipment includes a UP snow blower, a Union Pacific gas turbine locomotive, and the Southern Pacific 6916 Centennial locomotive that was used on the Bicentennial Freedom Train.

Restoration has begun on *Moon Glow*, but it has a long way to go. To make a contribution for the restoration of the car, send a check made out to "Union Station Foundation" to

Moon Glow *Restoration Fund*
Ogden Union Station—Utah State Railroad Museum
2501 Wall Avenue, Room 212
Ogden, Utah 84401

Please write in the check memo space "For *Moon Glow* restoration." Corporate donors are invited to contact Roberta Beverly at 801-393-9882

For information on the hours of operation of the museums, write to the address shown above, call 801-393-9886, or visit the Web site at www.theunionstation.org.

---

and there he put it on a set of trucks of an old Milwaukee Railroad dining car. However, the trucks proved to be the wrong kind, and the search began for a set that would fit the car.

In August 1987, Rawlins and his father went to Pocatello to put a coat of primer paint on the car and to cover the windows for transport and to hold down vandalism to the car. The car had been in good shape when Fernandez bought it, but it had been vandalized a few times over the years.

Meanwhile, plans were still being made to move the car from Pocatello to Ogden. Union Pacific was contracted to handle the move. They located a flat bed car in the system and had it moved to Pocatello. Dale Wood had found a set of trucks from an old railroad crew bunk car that would fit the *Moon Glow* and that were already stored in a gondola car.

Everything was in place in early 1988, so the car was loaded onto the flat bed car by Union Pacific and, along with the gondola car with the trucks, was placed in the consist of a freight train. Due to a tunnel at Bear River Gorge between Pocatello and Ogden, the two freight cars had to travel east to Kemmerer, Wyoming, from there to Granger, Wyoming, and then to Ogden.

Upon arrival at Ogden, the two cars were "stored temporarily on a rip track before being moved to the museum and put on the trucks."[7] The museum had arranged for Union Pacific to load the car and move it to Ogden, but the car was unloaded and placed on the trucks by local Union Pacific people at no cost. Now the search was on for a set of couplers to match the car, as well as parts to rebuild and refurbish the car to as near its original condition as possible.

Sometime around 1991, the *Moon Glow* was moved from the Union Station by rail to the Department of Defense's Army Rail Equipment and Diesel Generators facility next to Hill Air Force Base, where it was kept inside a building while some work was done on it. They discovered that Pullman-Standard had never put drains in the sills, and quite a bit of

While the *Moon Glow* was stored at Hill Air Force Base, the fluting was removed and a coat of gray primer was applied to prevent any further rusting. Now housed in a building at the Ogden Business Depot (formerly an army supply depot), the windows and seats have been removed as the car awaits restoration. Lee Witten Collection—All rights reserved.

rust had accumulated. While at the army facility, the car was sandblasted and a gray primer coat was applied.

In 2000, replacement glass for the windows was purchased but not installed. The Utah legislature granted $125,000 for all of this work, but when the money ran out, the army moved the *Moon Glow* outside. They tried to keep tarps over it, but that proved to be unsatisfactory.

In 2001, the car was moved to the Ogden Business Depot, a decommissioned army supply depot during World War II, between First and 12th Streets, where it is now housed in a former rail service building donated by the city.

In a real Cinderella story, the *Moon Glow*, or at least part of it, was transformed into a beautiful set for a film for the Sci Fi cable channel. Members of the production company came to Ogden to scout locations and visited Union Station. Bob Geier, then museum director, mentioned they had some cars in storage at the Ogden Business Depot. He took them out there, and the ones stored outside weren't what they were looking for. He mentioned that another car was stored in the building on-site. The *Moon Glow* was just what they were looking for, so they paid the Ogden Union Station Foundation $4,000 for the use of the car.

They cleaned out its filthy interior, set the windows in place temporarily, painted a part of the lounge area, and added carpeting, lights, furniture, partitions, and some fixtures. Filming was done in December 2004. When they were done, they removed the windows, furniture, lights, and fixtures, leaving behind the carpet, and the car was certainly cleaner than when they had found it.

Originally titled *Dead Rail*, the filmed aired in August 2005 as *Alien Express*, starring Lou Diamond Phillips, Barry Corbin, Steven Brand, and Todd Bridges. The film is about a senator, played by Corbin, who is running for president and traveling on a new super train heading to Las Vegas for a campaign stop. A meteorite crashes into a car near the tracks, causing the train crew to stop and survey the damage. The senator doesn't have time for this nonsense and insists that the train go ahead as planned. A policeman (Phillips) is called to the scene, where he discovers hungry aliens from the meteorite as well as a terrorist.

The *Moon Glow* and the *Train Of Tomorrow* have not only made railroad history but have also become a favorite among rail fans, historians, and everyone who had any contact with the train. It is comforting to know that such an important part of railroad history, as well as a part of the history of the United States, has been saved. Many people feel as Dan Kuhn did when he wrote, "It is my hope that someday I will have the privilege of riding in the restored *Moon Glow* coupled to the back of one of [Amtrak's] modern Superliner trains over the route to which it was originally assigned."[8]

The *Train Of Tomorrow* lives!

PART TWO

THE *Train Of Tomorrow* INSIDE AND OUT

The General Motors *Train Of Tomorrow* is the first new full dome train and includes an E7A passenger locomotive no. 765 and the 72-passenger chair car *Star Dust*, Pullman plan number 7555. The dining car *Sky View*, with a seating capacity of 52 in three dining rooms, is plan 7556. The sleeping car *Dream Cloud*, plan 4128, has two drawing rooms, three compartments, eight duplex roomettes, and 24 seats in the dome. *Moon Glow*, plan 7557, is the observation lounge car seating 68 people in the dome, the rear observation lounge, and two cocktail lounges, resulting in a total passenger capacity of 216 people. Part 2 is about this train, inside and out.

Built as Lot 6780 by Pullman-Standard Car Manufacturing Company and developed at a cost of $1 million to $1.5 million, the 411-foot train has many innovations. The most prominent features are the glass-enclosed domes, called Astra Domes and nicknamed the "greenhouses." The domes are 30 feet long and 10 feet wide, and they stand 2 feet above the standard height car roof, for a total height from rail to roof of the dome of 15 feet, 6 inches. The glass in the dome and throughout the rest of the train is also a new innovation, called Thermopane glass. Developed by Libbey-Owens-Ford, the Thermopane glass is heat-resistant, glare-proof, and tinted to help limit the amount of heat and infrared rays that pass into the car's interior. A pocket of dehydrated air between the plate-glass exterior and the shatterproof interior layer acts as insulation and as a barrier to exterior sounds.

Two other major changes are the use of outside swing hangers and rolling bearing journal boxes on trucks originally designed by EMD for use on locomotives and adapted for passenger car use. The innovative journal box design helps to cushion the lateral movement of the axles and, with the addition of special cushioning pads, dampens vibration.

Early truck design had the swing hangers mounted 56 inches apart on the inside of the truck frame, but on the *Train Of Tomorrow*, those swing hangers have been moved to the outside of the frame, 96 inches apart at the lower end. This design results in reduced body sway or roll on curves from 28 inches to 6 inches, and it also improves truck alignment and smoothes a rough roadbed.

Other ideas include diesel auxiliary power units for each car; an all-electric kitchen; curved surfaces to eliminate bumping into sharp corners; fluorescent light fixtures that provide bright, even lighting and eliminate dark corners; a public address system throughout the train that provides music from radios or a wire recorder; and a mobile telephone system that offers service within 25 miles of a major metropolitan area.

The cars are built of low-alloy, high-tensile steel of a welded-girder design with modifications made to accommodate the domes. Each car is 85 feet long, with an average lightweight of 150,000 pounds. The 71-foot locomotive has a loaded weight of 318,000, making the total lightweight of the train 920,000 pounds and the loaded weight approximately 977,000 pounds.

The train, including the locomotive, is painted blue-green, with ribbed stainless steel sheathing below the belt line and three stainless steel snap-on moldings in the letterboard.

The features common to all the cars are as follows: All the carpet is from Goodall Fabrics and is applied over a quarter-inch-thick felt pad. The carpet is of the "Seamloc" type and quality, with "Araby" colors (Jade, Araby Peach, Turquoise, Lido Sand, Persian Rose, Dove Taupe, and Silver Gray). Carpets applied over all passageway ramps are made into individual rugs and have an inlay design in a contrasting color.

New flooring and wall covering materials used throughout the cars—Es Es and V-Board—are plastic products manufactured by U.S. Rubber Company. Both are applied using special bonding cement. While Es Es material comes in a variety of colors (blue, green, red, brown, purple, coral, gray-blue, gray-green, and gray-tan), it can also be painted to match any part of the decor.

All of the paint used on the interior is semigloss by E. I. DuPont de Nemours. The exterior paint is the Dulux system also by DuPont, a high-gloss blue-green enamel. The underframe is painted with a black metal preservative finish.

The pull-down shades and hardware, all manufactured by the Adams and Westlake Company, are applied to all side windows, except the four curved windows in the rear of the observation lounge car, *Moon Glow*, and at the window in the rear door of the same car. There are no curtains in the Astra Domes on any of the cars. On the side facing out, the curtains are finished in a pebble grain blue-green Pantasote to match the exterior paint on the car bodies, except in the windows of rooms 1, 2, and 3 in the chair car and compartments A, B, and C in the sleeping car, which have a silver finish to match the stainless steel on the car's exterior. All of the curtains are lined on the inside with a natural colored satin glow material supplied by Goodall Fabrics. Unlike many other modern trains, there are no venetian blinds on the *Train Of Tomorrow*.

There are decorative but nonfunctional drapes hung throughout most of the train except for one set of functional draw drapes on the last four curved windows in the rear lounge area of *Moon Glow*. There are no drapes on any of the Astra Dome windows or on the door window at the rear of the observation lounge car. Goodall Fabrics supplied all the drapery fabric used on the train with the sateen backing material supplied by Lusskey, White, and Coolidge.

Haywood-Wakefield couch seats are used in the coach seating compartments and the lower-level rooms of the chair car, *Star Dust*, and in the Astra Domes in all the cars, with the exception of the dining car, *Sky View*, which has dining seats in the dome. The seats, which are of the rotating, reclining type, have a tubular metal construction with aisle pedestals and wall mountings. Among the features of the seats are sponge rubber construction in the seat and seat back cushions; high seat backs; adjustability to nine reclining positions; stainless steel kickplates; moveable footrests; and ash receivers in the armrests. There are no center armrests. The seats used in the Astra Domes have low backs, with the top of the headrest being only 25 inches above the seat cushion.

Before the construction of the *Train Of Tomorrow*, most of the lighting on trains was incandescent. But here the majority of the direct and indirect lighting is fluorescent, providing bright and even lighting throughout the interior. The fixtures, made by Luminator, are based on designs by the General Motors Styling Section. In what is believed to be a first in railroad passenger car construction, even the reading lights at the passenger seats in the chair car and the berth lights in the sleeping car are fluorescent. What few incandescent lighting fixtures remain are small and can be found in electric lockers, storage closets, and aisle lights throughout the train. Special lighting includes the steps from the passageway down to the private dining room on *Sky View* and compartments A, B, and C on the sleeping car, *Dream Cloud*; the "In Use" signs at the entrance doors of the toilets and dressing rooms in *Star Dust* and *Moon Glow*; edge-lit Lucite panels in the main dining room of *Sky View*; and an edge-lit sign in the passageway ceiling at the entrance to the upper cocktail lounge on *Moon Glow*.

Windows on the train are designed for maximum viewing of the passing scenery. Widths range from 3 feet in the upper and lower roomettes to 4 feet 7 inches in the compartments in the sleeping car *Dream Cloud*. The largest windows are 5 feet 2 inches wide in the two drawing rooms on *Dream Cloud* and in *Star Dust*'s three lower rooms and the forward and rear seating areas. The glass is Thermopane.

Clocks (32-volt) are mounted in the bolster on the forward dashboard in the domes of *Star Dust*, *Dream Cloud*, and *Moon Glow*.

The descriptions of the locomotive and the cars appear as chapters in the order in which the train has appeared on tour with both General Motors and Union Pacific. The individual descriptions of the interior design and decor begin from the front of the car (the A end) and proceed to the rear of the car (the B end) and include a summary of the carpeting and floor coverings, draperies, furnishings, colors, wall coverings, upholstery, facilities, and special features that distinguish each car. Most of the detail information about flooring, wall coverings, paint, curtains, finish, furniture, and draperies comes from the July 1946 specification book of the Pullman-Standard Car Manufacturing Company.

Come along on a tour of the *Train Of Tomorrow* . . .

CHAPTER 8

LOCOMOTIVE 765

Good morning. I'm Bob Stone, one of the assistant managers traveling with the *Train Of Tomorrow*. Our manager, Mr. Seward, told me you would be coming by for a tour. He wanted me to tell you that he's sorry he couldn't be here to take you around himself, but he's at a breakfast meeting with some local GM dealers talking about the train. We have quite a bit of time to tour the train before we open for public inspection today. Very few people get a chance to take a private tour. However, Mr. Seward told me I should give you the full 10-cent tour and not the nickel tour.

Let's start with the locomotive. This unit was given the number 765 because that was the EMD order number for its construction.

The *Train Of Tomorrow* has an E7A passenger 2-cycle diesel locomotive built by the Electro-Motive Division of General Motors. The way to tell the difference between freight and passenger units is that passenger locomotives have six-wheel trucks, while freight units have four-wheel trucks. Freight units are not equipped with boilers to produce steam heat in the winter, but in every other respect, the units are virtually the same.

This locomotive is 71 feet long with a loaded weight of 318,000 pounds. If you look closely at the locomotive's trucks, you'll see they have outside swing hangers, just like the passenger cars, which, along with a low center of gravity, contribute to a smooth and stable ride. Used for several years on EMD locomotives, the outside swing hangers reduce body sway on straight and curved track due to their placement 96 inches apart on the outside of the truck frame, rather than 56 inches apart on the inside of the truck frame.

The body's exterior is dressed up to match the rest of the sleek design used on the four domed cars. As you can see, here at the front is a large shield with GM and *Train Of Tomorrow* in polished stainless steel with a red painted background. On each side of the body is a raised star shield and insignia with the words *General Motors* within the star. From the star shield back to the rear of the locomotive's body is stainless steel fluting, matching the stainless steel fluted panels on the passenger cars. Near the top of the body, at the rear of the locomotive, are three stainless steel snap-on moldings to match the car letterboard moldings.

Walk around to the front and you can see that the front coupler is covered with a skirt that has doors providing access to the coupler.

Other than the cosmetics on the body, the locomotive is just like any other E7A unit coming off the assembly line at EMD. The philosophy behind this is to show railroad executives and officials, as well as the general public, that all EMD units are just as well

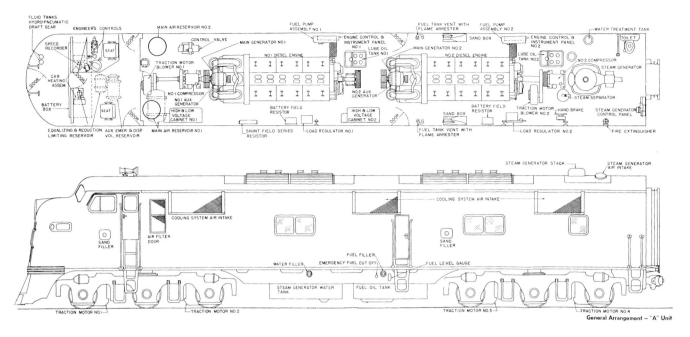

The floor plan of Engine 765, the E7 diesel passenger locomotive of the *Train Of Tomorrow*. GM.

Cover of the January 1979 issue of *Trains Magazine* with a drawing of the *Train Of Tomorrow*'s EMD E7 locomotive by Allen Brewster. Kalmbach Publishing Company—All rights reserved.

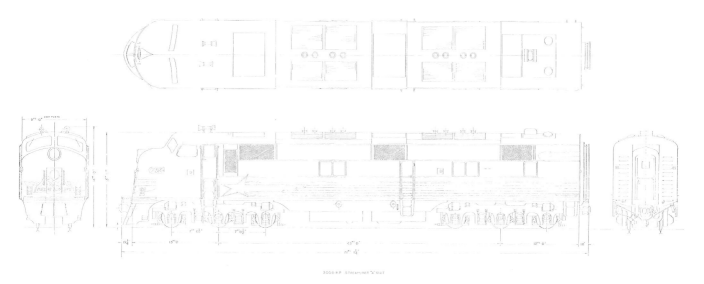

The left side, front, rear, and top of Engine 765. Notice the shield on the nose of the locomotive, as well as the shield and corrugated stainless steel on the side body. GM.

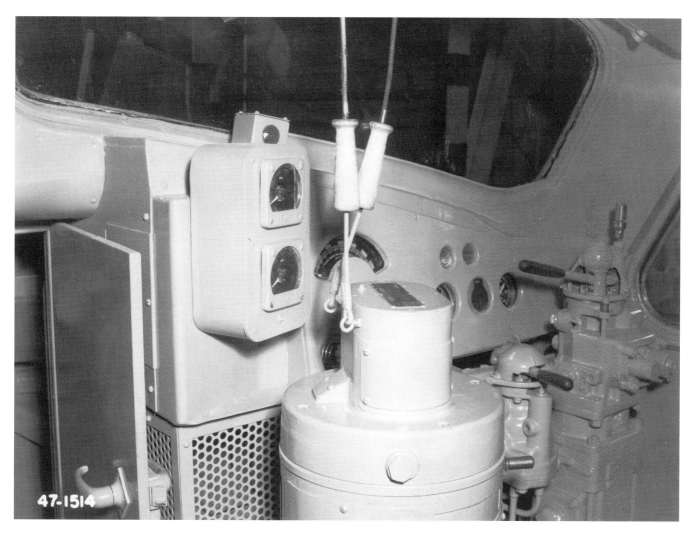

The engineer's controls of an E7A diesel locomotive like those on unit 765 of the *Train Of Tomorrow*. EMD—GM.

The left side of the engine room looking toward the rear of the locomotive. Diesel engine no. 2 is on the left. Just behind it is the traction motor blower, followed by the separator tank of the steam generator. GM.

The right side of the engine room looking forward. Diesel engine no. 2 is in the foreground on the left with the no. 1 engine at the front of the compartment. On the right in the foreground is engine no. 2's control and instrument panel. GM.

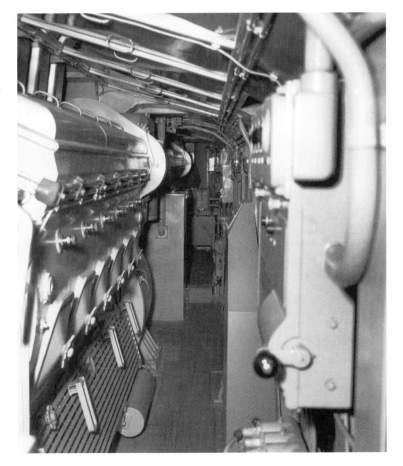

built as any locomotive used on exhibition. If special materials or enhancements had been used, then everyone would have a false impression of what a standard unit is like.

We are going on-board now. Watch your step. The staircases and ramps we use for tour guests entering and exiting the locomotive and cars were furnished by Aircraft Mechanics, Inc., in Colorado Springs, which makes stairs for airplanes.

As you can see, there are two swivel chairs in the cab, one for the engineer and one for the fireman. The controls on the modern diesel locomotives are similar to the controls of the old steam locomotive, so the engineers don't have difficulty making the transition from steam to diesel-electric power. The cab is soundproofed well enough so that the engineer and fireman are able to converse in normal tones. The cab is heated for winter and has roll-down windows for summer. The cab equipment includes warning bells and lights for the engines and boiler, automatic windshield wipers, and powerful defrosters.

Now if you will step back through this small door, we will go into the engine room. This locomotive has two 1,000 horsepower, 2-cycle, 12-cylinder v-type GM diesel engines. Power for diesel-electric locomotives is provided by means of an electric transmission, which consists of a direct current generator driven directly by each engine. This, in turn, powers the traction motors located on the trucks, which are geared directly to the axles. In all, there are two diesel engines, two direct current generators, and four traction motors in this locomotive.[1] The engines have a power range from 275 to 800 rpm. The power of the engines is controlled in turn by an electro-pneumatic governor that is controlled by the engineer's throttle. The throttle has eight notches, and every notch changes the engine speed by 75 rpm. To get the locomotive to move forward, current flows through the generator in the same direction as through the traction motors. To get the locomotive to go backward, the direction of the flow of the current is reversed through the traction motors.

To provide heat for the passenger cars, the locomotive is also equipped with a steam generator, which is located at the back of the unit behind the diesel engines.

Well, that's the locomotive. Let's go back into the engineer's compartment and down the stairs and take a look at the four cars.

CHAPTER 9

STAR DUST

L**ET'S GO LOOK AT THE WAY THE CARS ARE LAID OUT AND DECORATED. USUALLY** tours of the cars start back at the end of the train in the observation car, and people walk forward to exit at the front of the coach car behind the locomotive. But I'm going to start here, at the front, and we'll work our way back to the rear of the train.

The entire train and locomotive are painted a dark blue-green with a wide band of corrugated stainless steel below the windows and the molding strips at the roof line.

Star Dust is a 72-passenger chair car with seating on three levels, including the Astra Dome, and three semiprivate rooms below the dome compartment.

Traditionally, cars next to the locomotive have their dummy end forward, for the safety of the passengers, to prevent access to the locomotive through the door at the dummy end. *Star Dust* is designed the same way, with the dummy end forward. However, for exhibition purposes, GM has decided to carry the car vestibule end forward in order to provide an exit at the forward end of the train. That doesn't mean we aren't safety conscious. We certainly are. But it makes it easier to do it this way so we don't have to disconnect the locomotive from the rest of the train just to set up another set of stairs. Maybe I ought to explain what the term "dummy end" means. The dummy end of a car is the end without a vestibule, but it does have a door that allows passengers to move from one car to the next car. Why don't we take a look?

As you can see, the vestibule walls are painted gray, with the floor covered with gray-green Es Es material, with an inlay of green Es Es material. Es Es material is a new rubberlike plastic made by the U.S. Rubber Company. It comes in a variety of colors or can be painted. Just like all the cars that have a vestibule, the side vestibule doors are Dutch-type doors that open over the step trap. The end door leading to the car's interior is stainless steel and opens pneumatically. All you have to do is push on the door handle and the door opens by itself. Now we'll take a look at the inside. Step right through here.

Pretty nice, huh? As we walk to the other end, you'll notice that the car has two seating areas on the main floor in the forward and rear, coach seating in the Astra Dome, and seating for families or small groups in three semiprivate rooms in the area below the dome section.

The forward coach section seats 16, while the rear section seats 12. The Astra Dome has low-back reclining seats for 24 passengers. You may have noticed the layout of the three semiprivate rooms under the dome. Two rooms have seats for seven passengers, while the third room has two couches that will seat three each.

The right side of the chair car, *Star Dust*. UPRR.

This night shot shows how the semiprivate rooms fit nicely below the raised dome area. GM.

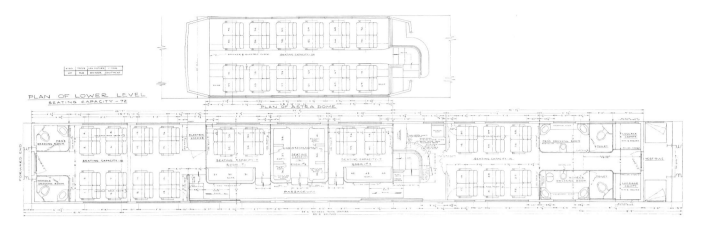

The floor plan of *Star Dust*. Pullman Technology, Inc.

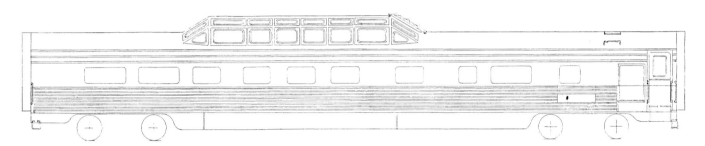

The left side of *Star Dust*. Pullman Technology, Inc.

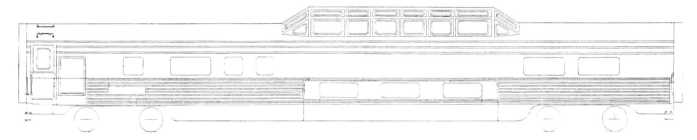

The right side of *Star Dust*. Pullman Technology, Inc.

Lighting throughout most of the train is fluorescent. The fixtures were supplied by Luminator, Inc. Fluorescent lights are certainly not new, but they are being used more often in all types of commercial applications. Our offices at GM are all lit with fluorescent lights. Even though they use less electricity, fluorescent fixtures produce bright, even light. Since less electricity is used, less heat is generated, so fluorescent lights are a lot cooler. Fluorescent lights allow designers more flexibility as well. As you can see here in the forward section, as in the rear seating section, there is direct fluorescent lighting above the center aisle, with fluorescent reading lights built into the luggage rack above the seats. One fixture is provided for each set of seats. When we get up into the dome, you will see fluorescent lighting fixtures above the center aisle as well. In the rooms below the dome,

The forward seating area looking toward the front of the car. Notice the two lavatories near the entrance door. UPRR.

the lighting fixtures are at the corner of the wall and ceiling, as is the lighting in the passageway under the dome between the forward and rear seating sections.

At the end of the car, you'll see two small lavatories just inside the door. The floor in the ladies' lavatory is covered in blue Es Es material, and the ceiling is painted cream. The wainscot is stainless steel. The inside upper wall and entrance door are painted a color called parchment, while the outside and aft walls are covered with satin finish V-Board covered in a wallpaper pattern called "Han-Tee Baroque." V-Board is a plastic wall covering material that, like Es Es material, is made by U.S. Rubber. It can either stand alone or can be covered with wallpaper. The men's lavatory is across the passageway. Its floor is covered in blue Es Es material, and the ceiling is painted blue. The lower walls are stainless steel, with the upper inside wall and the entrance door also painted blue. The outside and aft walls are covered in V-Board in a satin finish. The ceiling, walls, and doors in the passageway leading to the lavatories are Es Es material painted gray.

The aisle is covered in jade carpeting. Like all the carpeting in the train, there is a quarter-inch of felt padding underneath. The Seamloc carpet was made by Goodall Fabrics and of a level of quality they call Araby.

In the two main seating sections, the Es Es floor covering under the passenger seats is green with a red border. The ceiling, outside walls, and luggage racks are painted light yellow, with all the transverse, or bulkhead, walls done in gray. The wall below the window is covered with Es Es material painted yellow. The 16 seats in the forward section and the 12 in the rear section are adjustable to nine positions, with adjustable footrests and built-in ashtrays, and are covered in green "Sleepy Hollow" upholstery. The seat ends are painted yellow, except for three raised stainless steel horizontal stripes. The pedestals are painted gray. The seats also rotate not only to face the traveling direction but to face each other when families or groups are traveling together. With the four seats facing each other, a table can be put in so people can play cards or games. While people are walking through on tour, families will frequently sit down in facing sets of seats to try them out. The draperies at the windows have gray and yellow horizontal stripes with a latticework design at the top and bottom. The drapes do not close, so each window is equipped with an off-white pull-down shade. The outside of the shade is a dark blue-green to match the outside of the car.

If you'll walk back here with me, you'll see that from here the floor and ceiling slope down under the dome. The carpeting along the passageway under the dome is jade green. The two ramps have rugs with a silver-gray inlaid pattern. The ramps have moveable rugs because there are trap doors in the ramps for access to some of the equipment located just under the floor. As we walk down the ramp to the passageway along the outside wall, you

The passageway under the dome, looking to the rear of the car. Notice the zigzag pattern on the ramp rug. UPRR.

Looking into room 1, located under the dome. Notice that there are two facing sets of coach seats, plus a couch. These semiprivate rooms were designed to accommodate large families or small groups traveling together. GM.

can see that along the inside walls, leading to the three semiprivate rooms, there are shoulder-high partitions. The upper outside wall of the passageway is painted light yellow, as is the ceiling. The rest is painted gray.

The ceilings in the semiprivate rooms are painted light blue, and the carpeting is peach. From the light fixture at the corner of the ceiling and the wall to the window cappings, the outside wall is covered in Varlon wallpaper in a gray, orange, and peach checkered pattern. From the window cappings to the kick plates, the walls are painted gray. The

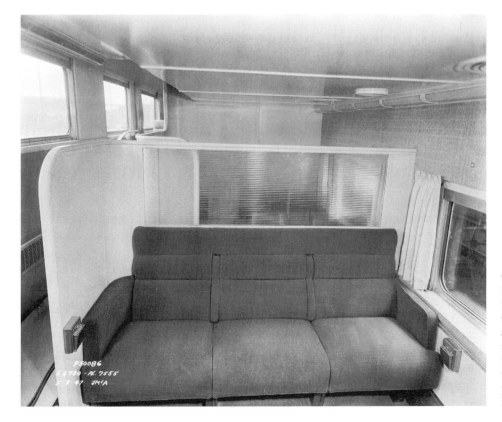

One of the two facing couches in room 2, which was designed to accommodate six people. The walls are only shoulder height, and partitions between the rooms have a ribbed glass in them to allow light to pass through but still allow for some privacy. Pullman Technology, Inc.

The facing coach seats in room 3. Pullman Technology, Inc.

inside of the partition walls are painted gray, with the finishing rails at the top of each partition painted light yellow. In the partitions between the rooms is Securit Linex Glass, which is a type of fluted or ribbed glass with the flutes running horizontally. You see glass like this used in partitions in office buildings. It is transparent enough to transmit light, but it distorts vision enough to provide some privacy.

The first and third rooms are identical in design and decor. The reclining passenger seats, mounted crosswise to the car, are covered in brown material, with the seat ends and pedestals painted gray. The couch, mounted lengthwise to the car, is also covered in brown upholstery.

Take a look into room 2. It has two facing couches mounted crosswise to the car that are covered in blue upholstery and have armrests separating each seat. Like facing sets for coach seats, these facing couches seem to be very popular with families. Of course, this is one way parents can keep an eye on what their children are doing.

The draperies in the three semiprivate rooms are beige with a pattern called Aristo. Like the windows in the coach seating areas, each window is equipped with an off-white pull-down shade, except this time the outside is silver to match the polished stainless steel molding that surrounds the windows.

As we move up the rear ramp from under the dome, in the area that includes the stairway to the dome, there is a water cooler, a locker for tables, an extension of the intratrain telephone, the public address system, chimes, a coat locker, and the porter's jump seat, which is covered in the same upholstery as the passenger seats. The ceiling in this areas is painted light yellow, and the walls are painted gray.

Now let's climb the stairs to the dome. Have you ever been in a dome compartment before? It's quite an experience. I still get a thrill every time I get the chance to ride in the dome, especially at sunset. It's beautiful. If you want to take a seat for a minute, I'll tell you about this room. The glass dome makes the roofline of the car only two feet higher than the standard car roof, with plenty of clearance in tunnels, bridges, and stations. The com-

The rear seating area looking forward to the stairs to the dome. The water dispenser is to the right of the staircase. GM.

partment is 30 feet long and 10 feet wide. The glass in the sides and the roof is called Thermopane, which is a new type of insulated glass that is basically like a sandwich—it has a pocket of air sealed between two pieces of glass. On the outside is a pane of heat-absorbing glass with a pane of laminated safety glass on the inside. The safety glass is tinted to reduce glare. The Chicago, Burlington & Quincy Railroad had a coach car redesigned as a dome car as a result of seeing the plans for the *Train Of Tomorrow*. However, this car is the first

Passenger seating in the dome. The seats are facing the rear of the car because the car ran rear end forward during tours by General Motors and Union Pacific Railroad. This put the vestibule at the front to provide an exit for people taking a walking tour through the train. The dial on the leather dashboard is a clock. GM.

newly constructed dome car, as are the other three on this train. The response on the Burlington car was overwhelming and gave those of us at GM the idea that we were on the right track with the *Train Of Tomorrow*.

Taking a look at the decor in here, you'll notice the jade carpet running the length of the dome, including under the seats. The low-backed reclining seats are covered in rose-colored material, with the seat ends and pedestals painted tan, as are the moldings, window frames, and walls. The seats are low-back in the dome so that people can see better. If the seat backs were high, like the ones in the coach sections, the passengers wouldn't be able to see forward. Look up to the front of the dome. There is a dashboard bolster covered in gray-green London Glaze leather. The bolster continues along the sides of the car directly below the side windows, from the dashboard to the rear partition. Mounted in the dashboard is a clock. Some people have said there should be a speedometer mounted there. The ride on the *Train Of Tomorrow* is so smooth that people are astounded to discover they are traveling 100 miles per hour.

At the tail end of the car is the rear coach seating section, decorated identically to the forward section, except there is seating for 12 instead of 16. Behind this are the ladies' dressing room and lavatory on one side and the men's accommodations across the pas-

The lavatory and dressing room in the men's lavatory at the rear of the car. The small sink on the left is known as a dental lavatory. UPRR.

sageway. Let's walk to the back and take a look. Step through the door on the right. In both the ladies' lavatory and dressing room the floors are covered in red Es Es material. The ceilings are painted light yellow. The lower walls of the dressing room are stainless steel, and the upper walls and the doors are painted gray. The draperies are a pattern called Moon Rhythm Gray, which has large yellow circles against a gray background. The vanity chair is covered in blue-gray nylon upholstery. The dressing room has two washstands and a dental lavatory. In the lavatory the lower walls are also stainless steel, with the upper walls, door, and ceiling painted gray.

Now we will move across the passageway and take a look at the men's dressing room and lavatory. Like the ladies' compartment, the floors of both rooms are covered in red Es Es material. The ceiling, doors, rear wall, and outside wall are painted in gray-blue. The forward wall and passageway wall of the dressing room are covered with V-Board and wallpaper called Rendezvous, sort of a Gay Nineties theme. I think the wallpaper with the figures of the men and women on it is an odd thing to put in a room like this. It looks more like cartoons. But I guess the designers know best what will look good to the passengers. The drapes are rust-colored. There are two washstands and a dental lavatory. The walls of the toilet room have stainless steel from the floor to the wainscot, with gray-green paint on the upper walls, the ceiling, and the door.

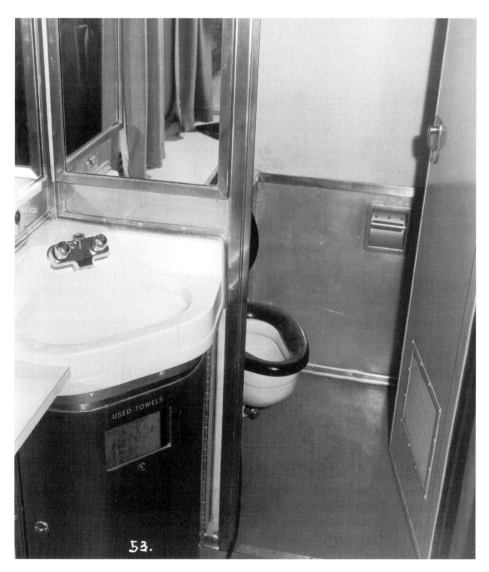

One of the sinks and the toilet in the men's lavatory. UPRR.

The door leading from the vestibule into the rear of the car. Notice the luggage compartments on either side of the passageway, one open and the other with the door rolled down. These luggage compartments also had doors to load suitcases and bags from the outside. UPRR.

You'll see luggage compartments on both sides of the vestibule door. Both compartments have rolling doors on the passageway side. There are doors on the exterior of the car that give access to each of the compartments for the quick and easy handling of passengers' hand luggage. The walls of the interiors of the compartments are painted light yellow with the shelves covered in gray Es Es material. The floors are covered with green Es Es.

In the rear passageway you'll see that the walls, ceiling, and door casings as well as the doors to the dressing rooms and rolling doors on the luggage compartments are painted gray. The door leading to the vestibule, like the door at the dummy end, is painted green.

All four cars have 10-ton air-conditioning split into 4-ton and 6-ton units. Underneath each car is a diesel generating unit that supplies electricity. All of the cars can be run together, or they can actually be set off and still have power available on-board. The units are so powerful that the power unit from one car can supply the power for another car should a breakdown occur. The power unit on each car is large enough to provide power for several homes. It has been said that if you combine the power from the four cars, as well as the massive generator for the kitchen equipment, there would be enough energy generated to run a small town.

Well, that brings us back to where we started. Let's walk back through the car and go see the dining car.

CHAPTER 10

SKY VIEW

This is the dining car, *Sky View*. It has three levels and seats 52. The main dining room seats 24, the dome seats 18, and the reserve private dining room in the space below the dome section seats 10. This is the first dome diner to be built, and General Motors decided to do it up right by building the first diner to have an all-electric kitchen.

We'll start here with the passageway. To the right is the crew's toilet. There the floor is green Es Es material, just like it is here in the passageway. The wall from the floor to the wainscot is stainless steel. The walls from there up, as well as the ceiling and the door, are painted green. There is a toilet and a washbasin. The end wall and the wall in the passageway from the end door around the crew's toilet to the main passageway are painted red. The ceiling in the end passageway and the wall on the other side of the passageway are painted green. Opposite the crew's toilet is some of the refrigeration equipment. The unit closest to the end door is the icemaker. It is capable of producing 225 pounds of ice per day, more than enough for the train's needs. The empty space next to it is for a unit for food storage that is accessible from inside the kitchen. The on-board refrigeration equipment is mechanical and has a total capacity of 111 cubic feet, which is equal to 15 standard home refrigerators. The refrigeration fixtures are Pullman-Standard built with the condensing units, motors, fan and pulley assemblies, belts, evaporators, tubing, fittings, refrigerant, and antifreeze from the Frigidaire Division of General Motors.

Straight ahead is the entrance to the kitchen, which covers the forward third of the car, while the connecting pantry shares the space under the dome with the private dining room. I love coming here because it always smells so good. Our head chef, Joe Schneiderbauer, and his crew turn out fabulous meals as well as hors d'oeuvres for receptions. Sometimes, when we are en route, the kitchen crew will prepare meals for us. I particularly enjoy eating breakfast in the dome. It's a great way to start the day. The only time we didn't get meal service was when we were at the Chicago Railroad Fair. We had to make arrangements to get our own meals until it was discovered that the choices were limited and not very good at that. So GM decided to provide sandwiches and snacks to the fair crew. I have seen the kitchen crew come up with enough food in this limited space to feed 800 people at a reception.

I don't know all the technical information about the kitchen, pantry, and dining rooms, so I've asked a couple of people for their help. This morning, while they fixed breakfast for

The right side of the dining car *Sky View*. The large blank area is the kitchen and pantry. The compartment for the large generator to power the all-electric kitchen is at the left of the photograph. UPRR.

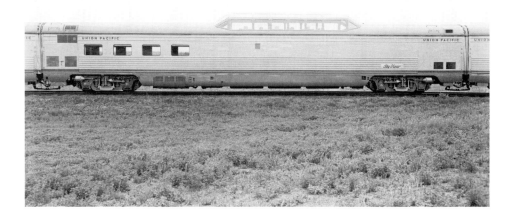

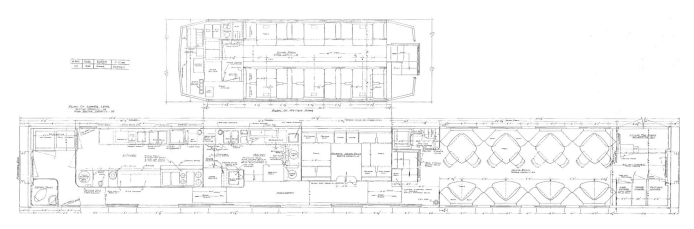

The floor plan of *Sky View*. Pullman Technology, Inc.

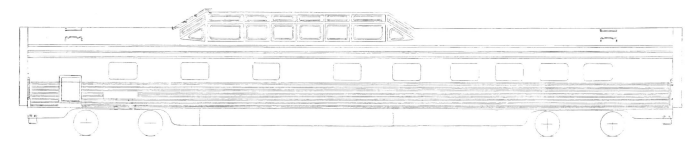

The left side of *Sky View*. Pullman Technology, Inc.

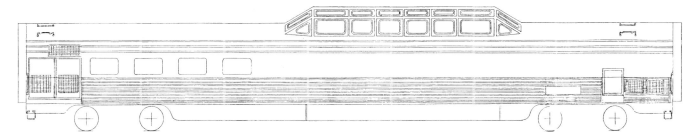

The right side of *Sky View*.
Pullman Technology, Inc.

the entire crew, I mentioned that I had a VIP guest coming. Buddy Tyson, our head cook, said he will show you the kitchen and the pantry, and William Cruickshank, the assistant steward, said he will show you around the dining rooms. All the kitchen employees, the stewards, the waiters, and the car porters are Pullman employees. They are supervised by a group of people from Pullman called inspectors. All the people from Pullman are very experienced, and some have been with the train since the beginning of the tour. Others have joined along the way. All of them were handpicked because of their ability to handle

The ice maker in the forward passageway. UPRR.

special assignments and their professionalism. It seems as though they always know the right thing to do or say.

I have some things I need to do to prepare for today's exhibition, so I will go find Buddy for you.

Hi, my name's Buddy Tyson. I'm the head cook. I told Bob that I'd show ya around the kitchen and pantry. First, let me tell ya jest a bit about myself. I was born in Atlanta and went to work as a cook with Pullman 27 years ago. I've worked on trains all over the country, like the *20th Century Limited* and all the Florida trains. I've also worked with Joe Schneiderbauer on the president's car. Besides that, I've worked on the private railroad cars of a lotta famous people. I've been on the *Train Of Tomorrow* since a couple of days before the preview run made in May of 1947.

If y'all follow me, I'll take ya on a tour. The kitchen is all stainless steel and equipped to handle jest about anything that might come up. Step up to the door here. Y'all see how the floor looks like a pan of metal welded together to form one piece. It banks up the wall slightly and then slopes to a drain in the center. The metal is covered with Martex, which they say is supposed to be an anti-slip material, but if ya git a lotta water on the floor or if some food's spilled, it can be tricky. Y'all still have to watch your step.

We're gonna go inside to have a closer look at what's there, but we'll have to stay outta the way of the cooks. People on walk-through tours don't git to come in here. They jest take a look through the door. I'm gonna start on the right at the front and work my way round the kitchen back to where we are here at the door. Then we'll go down the hall and I'll give ya a look at the pantry.

The first thing we come to is the hot water heater with the water tank in the ceiling above us. This is just for the kitchen. Next are the pastry board and Kitchen Aid mixer. All the cakes, pies, and pastries used on the train are made here. Next to that is the Hotpoint

The kitchen looking toward the rear of the car and into the pantry. This was the first all-electric railroad kitchen. In the foreground on the right is the range with the broiler above and ovens below. The dishwasher is on the left with the rounded top. UPRR.

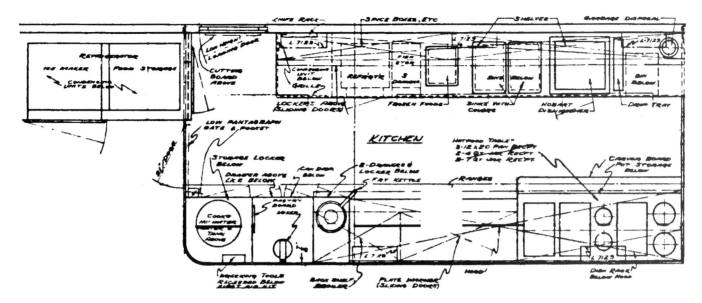

A detailed drawing showing the layout of the all-electric kitchen, located at the front of the dining car. Pullman Technology, Inc.

fry kettle. Everything in the kitchen is electric, so we don't have to worry about charcoal or gas. It's a real pleasure to work with this stuff. It's so much easier and it don't make as much of a mess as charcoal or gas. For example, with charcoal we weren't able to have fry kettles like this. We had to fry everything in skillets. It was such a mess, and it tied up one of the burners on the stove, but with this you can be cooking something over here and still have all the burners on the stove open. I'm gonna miss all this when the tour is over. I jest hope they build some more cars with kitchens like this one.

Over here are the three General Electric ranges. Above the first one is the broiler, also made by GE. Each of the stoves has an oven below, and above the other two stoves are plate warmers. Notice that the stoves don't have burners like you see on a stove at home or in most restaurants. It's just a flat surface that you can put anything on top of, or you can use it as a griddle.

From the end of the ranges to the wall that connects to the pantry is what we call a hot table or steam table, where we can keep things like soups and gravies hot in pots and meats or vegetables hot in pans. The fixture itself was made by Pullman, and Hotpoint made the electrical parts. We put water down in the bottom compartment and heat it, and then the steam keeps the food hot in the pots and pans without causing it to burn or stick. There is room for three pans or half pans, two 4-quart pots, and two 7-quart pots. On the front of the steam table is a carving board. Above the steam table is storage for dishes and pots. There is also some storage under the steam table. Above the whole right side is a hood to pull out the steam and smoke.

As we turn around to go back, you'll see that a part of the dome section cuts into the kitchen. Actually it's part of the waiter's section just above us. It's where the sink is. If you look in the corner next to the wall, you'll see the GE garbage disposal, and next to that is the Hobart dishwasher with storage for clean dishes above it. Then we come to a sink that is used not only for cooking but for cleaning pots and pans as well. Next to the sinks is a frozen food locker with a cover. All the refrigerators were made by Pullman-Standard, but the mechanics were made by Frigidaire.

Then we have a long open counter space where we can do some prep work. We can get a little more space by putting covers on the sinks and the frozen food locker. Below this work area is a refrigerator and a place to store fish in three compartments. Above here are the spice boxes and some storage.

This is the view of the kitchen looking toward the front of the car. The door straight ahead leads to the forward passageway. To the right is a freezer and next to that is an exterior door for loading supplies. UPRR.

At the end of the counter is a low door that's used for loading supplies. There is also a little fold-down seat for us if we git a break in our work. Finally, there's a big food storage locker.

Now, if you'll follow me down the hall, I'll show you the pantry. You'll notice how the floor and the ceiling slope down as we walk along. That's where the dome section is, and the floor and ceiling are depressed so we can use the area below the dome. I think it's pretty clever, but I'm not one of them engineer fellows, so I can't tell ya anything more about it. I'm sure Bob knows all about it. Ya can ask him if he hasn't already said something.

The pantry and a small private dining room are under the dome. The pantry is where the salads and desserts are made and where waiters pick up their orders. They git their salads, desserts and some of the drinks, like coffee, tea and milk from here. As we go through the door, take a couple to steps down and look to the left and you'll see the dumbwaiter that's used to send food up to the dome. Next to that is the opening that leads to the kitchen. Food going up to the dome can be put in the dumbwaiter from either the kitchen or the pantry. A kitchen helper called a pantryman works down here to help the waiters and to help handle the orders for the dome. On the wall next to the dumbwaiter y'all see a thing that looks like a telephone. The waiter in the dome can talk to the pantryman over that to give him the orders for the kitchen or to ask for something else he might need from the pantry. Below the telephone is a large wheel that is turned to crank the dumbwaiter up to the dome and back down again. Next to the dumbwaiter, right here next to the door, is a place for ice storage with a glass rack above it.

Two cooks prepare meals on the right as a waiter does dishes on the left. The cook in the foreground is Joe Schneiderbauer, the Pullman inspector of cuisine on the General Motors 28-month tour. GM.

Now we'll turn right and work our way round the pantry. At first we see the service shelf with the pie locker on top, silverware drawers below that, and a cup warmer at the bottom. Next to that are the two 2½ gallon coffee urns. Below the first one is some tray storage, and below the second one is a refrigerator. Over there in the far corner is an ice cream storage locker. Next to that is some open counter space with a set of refrigerators below. Above the

A cook passes a cup of soup to a waiter in the pantry. GM.

The rear of the pantry looking toward the rear of the car. To the right is a dessert locker. At the rear are two coffee urns. On the left is the area for making salads and sandwiches. UPRR.

counter space are the bread locker and the cold pan, where we store lettuce and vegetables for salad or sandwich fixin's. The pantryman can put together salads or sandwiches for the waiters without having to ask the cooks for anything. At the end of the open counter is a juice extractor. Then there is a Hobart glass washer, a set of sinks, and some more counter space with another GE garbage disposal in the corner there. Wherever there's any open space, it's used for storage of one kind or another.

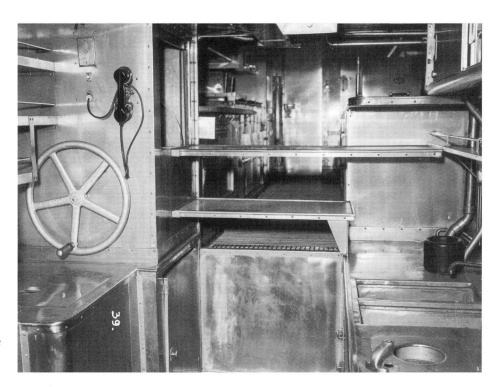

Looking from the pantry into the kitchen. To the left of the top shelf is the dumbwaiter. UPRR.

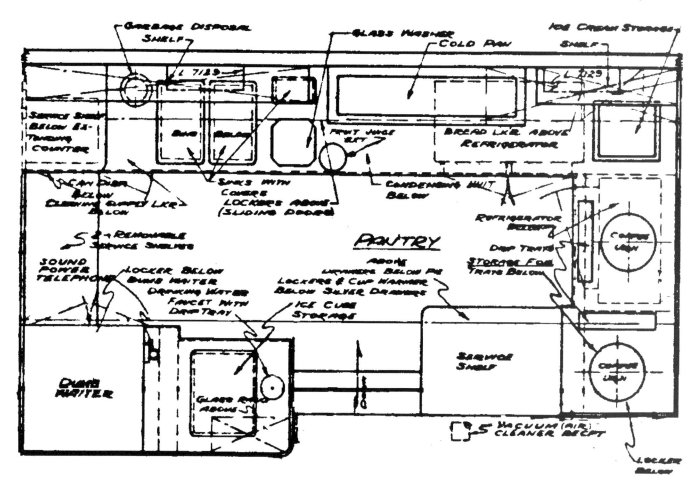

An enlargement of the pantry area, located under the dome, just to the rear of the kitchen. Pullman Technology, Inc.

That brings us back to where we started. I think I've showed you jest about everything. I can't think of anything else. I know we're gonna serve lunch to some people today before we open at one o'clock. I hope y'all git a chance to join us. If they don't invite ya, y'all jest come by the kitchen here and I'll make sure ya git fed good. I see that Mr. Cruickshank is standing in the hall waiting for ya. I gotta git back to work anyway. It's been my pleasure, and like I said, if they don't invite ya to lunch, y'all come and see me.

Good morning, my name is William Cruickshank. Mr. Stone asked if I would show you around the dining rooms when Mr. Tyson was finished. I hope he showed you everything you wanted to see. I can tell you this: he's a wonderful cook. I cannot remember any meal he has prepared that I didn't enjoy. He is famous for his barbecued ribs and his fried chicken.

Mr. Stone said I should tell you something about myself. I have been with the Pullman Company for only 12 years, but I have done very well in that time. I was hired into an office position at Pullman in Chicago in the commissary department, but I didn't stay there very long because I was asked to fill in for an inspector on a special assignment one weekend, and I have been in the field ever since. I am probably the only steward in the company today that didn't start in the kitchen or as a waiter. I guess because I have some formal training in hotel and food services I was made an assistant steward right away, and I have worked mostly on the trains going west out of Chicago to either Los Angeles or San Francisco. I have had the pleasure of serving many well-known people, sometimes more than once. When we had our run for celebrities in Los Angeles, several of the people, such as Walter Pidgeon and Betty Grable, remembered who I am and called to me by name. That made me feel good. While I am usually a steward in regular service, I was willing to sign on for this assignment as an assistant, just for the experience. I have been with the *Train Of Tomorrow* since it left Chicago on its preview run to French Lick, Indiana, on May 26, 1947.

Mr. Stone said I should show you everything and not leave out the smallest detail. Since we are standing here at the entrance to the pantry, why don't we start with the passageway? The walls and ceiling, from the crew's toilet to the base of the staircase leading to the Astra Dome, are painted green and accented with turquoise carpet. The inlaid design in the ramp rugs is peach.

Right next to the pantry in the section under the dome is a small dining area called the reserve private dining room. It is meant for small parties, groups, or families to dine together. You'll notice that the step leading down to the room has a glass step tread lit from below with a small reminder to *Watch Your Step* placed on the glass. Lighting for this room is supplied by a fixture at the corner of the wall and the ceiling on the walls that run crosswise of the car. As you can see, the carpet is turquoise and the rest of the room is done in various shades of red. The wallpaper, which runs from the ceiling to the tops of the seat backs and the level of the table tops, is a Varlon paper with a pattern of small red, pink, and blue flowers with green leaves printed on a white background. Varlon wallpaper has a clear plastic coating that makes it easy to wipe off any stains. It is used in many high traffic areas on the train. The ceiling is painted ivory, and the finishing rail at the opening to the room is painted parchment. If you will look under one of the tables, you see that there is coral super needlepoint fabric that runs from the bottoms of the tables to a line even with the bottom of the seat cushions. From there down to the floor, the walls are painted gray.

The seating arrangement in this room is unique. There are two tables set up like booths with benches or sofas on both sides of a table. There are two such set-ups in the room. The outside seat on both sides of each table fold up to allow easier access to the inside seat. Then there are two loose chairs stored against the outside wall that can be pulled around to the end of each table, giving you five places at each table. There are headrest bolsters that are placed above all the built-in seating. The backs of the loose chairs fit flush with the bolster, so the room has a nice finished look when the chairs are not being used. All of the seating surfaces are covered in the same coral super needlepoint fabric that is on the walls beneath the tables. Just like all the tables in the dining car, the tables have Formica on the top and sides.

So you can see it's easy to accommodate two groups of four or five. The only complaint we get about this room is the lack of a window, but I personally believe not having the window adds to the intimacy of the room.

Why don't we take a look at the dining room in the dome next? We'll go back out into the passageway and turn left and make another left to the staircase. The area around the steps and all the way around to the other side of the steward's room is covered in oak Flexwood. This is one of the only walls on the train that is covered in wood, and I think it adds a nice clubby yet elegant touch to the car. Now if you will join me, we will climb these steps up to the dome. Please watch your step.

I think this is a most impressive space and certainly a favorite of everyone who sees it. This is the first dining car to be built with a dome. Judging by the response to this compartment, this won't be the only one built. A lot of people have been served in this area, and everyone had something positive to say. When the cooks prepare a meal for the staff, almost everyone comes here to eat. The unfortunate part is that there are only enough seats for 18 people. That is enough seating for the staff, but when you have a large group

Four people dine in the private dining room under the dome. The dining room could seat up to 10 people at two tables. GM.

of important guests or members of the press on-board, it is difficult to handle the requests for seating in the dome. Sometimes people get angry with us when they cannot be seated up here, but we have a set policy to help guide us, and we can always refer to that. Usually, we know ahead of time who is going to be on-board for the day, and we frequently receive a list of those who are going to be invited to dine in the dome. That is another way that helps us deal with the people who get upset because they cannot get into the dome. We just tell them the people there were invited. When the meal service is finished, we then allow anyone who wishes to come up here to have a look around. As a matter of fact, when there is no food service, we encourage people to come up here. Sometimes they are afraid they will get in trouble for being up here. We just make sure everything like china, glassware, flatware, and hollowware are secured so no one will be tempted to take anything. Of course, General Motors wants us to encourage people to see all parts of the train so that the public will begin to make demands of the railroads for new equipment like this, built with General Motors parts.

We are only going to have 16 people for lunch today, so everyone will be able to have a seat up here. While the tables are set for lunch now, all of this will be cleared away before walk-through tours begin. Today's menu is cream of vegetable soup, followed by Chicken Cordon Bleu with seasoned rice and asparagus. For dessert we will serve Cherries Jubilee. We are planning to serve a light white wine with the entrée, and coffee, tea, and milk will also be available. A waiter from the lounge car will serve drinks before the meal is served and later help with serving the wine. We have it down pretty well now and can serve and clear a meal in an hour without making the guests feel they are being rushed. One interesting point about the food service is that when the staff has a meal together, we lay out all the china, linen, and other implements for ourselves. We don't get to have that many meals on-board, but we want to make it a memorable experience for all of us. As a matter of fact, both GM and the Pullman Company seem to encourage this as a little extra perk for all our hard work.

As you will note, the aisle up here is off center allowing for tables for two on one side and four on the other side. It also allows for a slightly wider aisle for the waiters to move around without bumping into people. Seating up here is also in booths. It would be too cumbersome to have tables and chairs. Besides, design considerations don't allow for that type of set-up.

The color of the carpet here and on the steps is called Lido Sand. The walls and window frames are painted gray-blue. The seat cushions are brown leather, while the back cushions are covered with a sandalwood cloth. At the tables that seat four people, the outside seat folds up to allow easier access to the inside seat, just like the booths in the reserve private dining room. Each table has a built-in light at the center of the table against the outside wall. There is also overhead fluorescent lighting from fixtures located in the ceil-

The dining area in the dome looking toward the rear of the car. Tables on one side could seat four people while those across the aisle were built for couples. This provided a wider aisle for the waiters to maneuver heavy trays of food. UPRR.

ing of the center aisle. Let's take a walk down to the forward end of the car, and I'll show you the waiter's station.

The floor in this area is not covered with carpet but has brown Es Es material instead. The work area is fairly self-contained, so the waiter has some independence from the kitchen. On the right is an area that includes space for serving tools, a cash drawer, and a storage space for miscellaneous items. Below that is a storage cabinet for clean linens. A little further forward is a cabinet that has two holes in the counter top. One hole leads to a container for soiled linen, and the other is a garbage chute. This helps hold down on waste odors. Up near the front of the car are two sinks. Moving around to the left, there is a lock box that is the silverware rack. Next to that is an electric four-slice Toastmaster toaster. Below the silverware box and toaster is a refrigerator. Farther to the left at the back of the counter space is a Cory coffee warmer. Just in front of that is the opening for the dumbwaiter that brings up food and supplies from the kitchen and pantry just below here. Down there in the corner, next to the dumbwaiter, is the telephone connected to the kitchen or pantry for orders or requests for supplies. Over there in the far left corner is a storage space for water bottles. Turning the corner, there is the ice well. Next to that is open counter space and a carving board for placing trays of food from the dumbwaiter or as a work area for any food preparation the waiter may have to do. Finally, below the open counter is the china locker. On the back of the last booth there is a holder for trays and menus.

Now we will move back down to the main dining room. As we come down the stairs, you will notice a plant at the curve of the staircase. Each of the staircases has a plant al-

Looking toward the front of the dome dining area, showing the waiter's station. The dumbwaiter to the pantry is on the left under a set of doors on the top of the counter. The station is equipped with a toaster, coffee warmer, sink, refrigerator, and ice well. Pullman Technology, Inc.

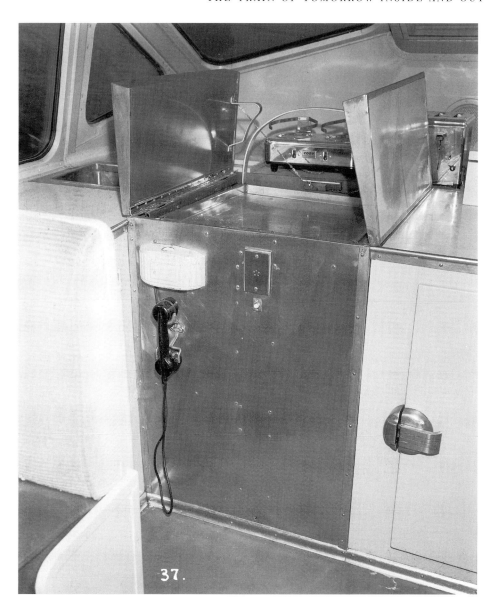

The dumbwaiter and intercom in the waiter's station in the dome. The doorbell just above the intercom was used to get the waiter's attention. UPRR.

cove like this for a plant or flowers. The flowerpot is stainless steel. The pot has a perforated bottom, and the well the pot sits in has a small drain. In the cabinet below the plant alcove is liquor and bottle storage.

At the base of the staircase is a small trap door that leads to storage where we keep extra cases of beer before we put them into the refrigerator. Over against the outside wall is the steward's room. Actually, it's more like a closet, but we really don't need any more space than this. As a matter of fact, this is one of the few cars on the rails that have a separate section like this. Usually I have to work out of a corner of the kitchen, or sometimes there is a built-in credenza that is set aside for the steward. First, you will note that the exterior wall is covered in the same Flexwood that surrounds the dome's staircase. Now, if you'll step up to the door I'll show you how well this is laid out. Someone gave this a lot of thought. Along the wall on the right is a locker for our personal belongings and our uniforms as well as the supplies we need. In the right corner is the desk with a light mounted in the wall above it. There are two drawers and a small storage cabinet below. Above the desk is a zinc-lined cabinet that is used as a cigar humidor and for cigarette storage. On the left wall is a double-door refrigerator that will hold up to eight cases of wine, beer, and

Actress Alice Faye and her husband, bandleader Phil Harris, enjoy dinner in the dome. GM.

The steward's compartment at the bottom of the staircase to the dome. UPRR.

The main dining room from the stairs to the dome looking toward the rear of the car. GM.

soda pop. Mounted on the wall between the desk and the refrigerator is one of the stations of the intratrain telephone system. The floor is covered with brown Es Es material, and the walls and ceiling are painted a beautiful shade of blue. It is a real pleasure to have this workspace. I love it.

Now we'll take a look at the main dining room. I think a lot of thought went into this room. People really enjoy dining here. First, you will notice that like the dome, the aisle is off center, with tables on one side for couples and tables on the other side for parties of four people. The tables for two are triangular, with the point facing the outside wall. The seats are built-in so that the people are sitting at an angle, facing each other and the aisle. The tables across the aisle are placed cornerwise to the car with built-in seats along the outside wall, just like the ones on the other side, plus two chairs on the aisle. This arrangement allows people to be seated without disturbing other people, plus it gives a lot more aisle space for the movement of the waiters and stewards as well as the customers. The tabletops are mounted on curving pedestals attached under the seat cushions of the booth. Between the booths and along the wall are small triangular platforms that can be used for holding menus or condiments. We use them to display floral arrangements.

There are two types of lighting in here. First, there are fluorescent fixtures above the aisle, but there is also indirect lighting from fixtures located behind aluminum-clad valances just above the windows. When we turn off the overhead lights, the indirect light-

People having dinner in the main dining room. This photograph was shot looking toward the front of the car. The steward's compartment is on the right at the base of the stairs to the dome. GM.

ing provides a nice soft light that makes for a pleasurable dining atmosphere. One thing that helps contribute to that, though, is the fact that the walls and ceiling are painted parchment. Add the fact that the carpet is a soft shade of peach, and you have a room that makes the customer feel very warm and comfortable. The seats of the booths and the eight chairs are covered in Chartreuse super needlepoint. The draperies carry out the Chartreuse with blue added on a beige background with a pattern called Acouthus Pistachi. Besides the draperies there are off-white pull-down shades. The bases of the booths are covered with the oak Flexwood. At the other end of the room is another wall covered with oak Flexwood. The center of each wall has an edge-lit Lucite panel with the face of a woman etched into it with a dark blue background to match the blue in the drapes. To make up for the difference in the width of the end walls, due to the difference in the size of the tables on both sides of the aisle, the wall on the side with tables for four has a small partition of polished plate glass with rounded edges to prevent anyone from getting hurt should they bump into it.

Beyond the end wall is the passageway that leads to the rear end of the car. On one side is the compartment for the 40-kilowatt generator and its control panel needed to supply electricity for the electrical appliances on-board. Across the aisle, from the wall at the end of the dining room to the wall at the end of the car, is a linen locker, followed by a crew locker and the electrical locker. The floor in the passageway is covered with brown Es Es material. The walls and ceiling are painted tan, and the door at the end of the car is blue.

A photocartoon from the September 1978 issue of *Trains*. Kalmbach Publishing Company—All rights reserved.

 Do you have any questions? I hope I have been able to show you what you wanted to see in this car. I see Mr. Stone is coming through the passageway under the dome. If I can be of any other service to you, please let me know.

 Hi. I hope you had a good time with Mr. Tyson and Mr. Cruickshank. I knew when I left that you would learn more about this car from them than if I had taken you through. Well, let's go through the door at the end of the passageway here, and we'll take a look at the sleeping car.

CHAPTER 11

DREAM CLOUD

T HE SLEEPER, *DREAM CLOUD,* HAS TWO DRAWING ROOMS, THREE COMPARTMENTS in the space below the dome, and eight duplex roomettes. With all the beds filled, the car will sleep 20 passengers. The way the rooms are designed, there's 50 percent more floor space for daytime use, and yet there is maximum space for sleeping arrangements. All of the beds are mounted lengthwise to the car. In the drawing rooms and compartments, this affords a view directly out the large windows from the sofa. Besides, it's known that sleeping is more comfortable when the beds are positioned this way in a sleeping car. In the dome, there's a reserved seat for everyone booked on the sleeping car. The drawing rooms and compartments are all decorated in different colors and fabrics, so no two rooms are the same. All the upper roomettes have one color scheme, while the lower roomettes have another.

Mr. Hill, your tour guide, if you could call him that, has been with the Pullman Company for more than 35 years. He has worked for a lot of very important people, including the president of the United States on his private car, as well as many of the presidents of the railroads. He probably has more seniority than any other car porter in the company, and he does a lot of the training of porters for special movements and events. Because he was on special assignment when the tour of the *Train Of Tomorrow* started, he was not able to join us until about two months later. Before coming on-board he went to the Pullman-Standard plant in Chicago to learn all he could about the car, so you'll probably learn more from him about *Dream Cloud* than I could tell you. I'll go tell him you're ready, and I'll see you later.

Morning. I'm Charles Hill, but you can call me Hill. Everybody else does. Robert Stone asked me to show you through the car. I'm very proud of this car. To tell you the truth, it is probably the nicest car I've ever worked on. It's even nicer than the car used by the president of the United States. As a matter of fact, that car's rather plain next to this one. Inspector Schneiderbauer, the head chef, Mr. Tyson, the head cook, and I have all worked on that car, and of course it was a privilege. But we all agree this is a lot more luxurious and just about as much of an honor.

Let's start here in the front of the car. To the right of the door is a general use toilet. If you'll open the door there you'll see that the floor is purple Es Es material. There is stain-

The right side of the sleeper *Dream Cloud*. At the left are the windows of the duplex roomettes. Under the dome are the three compartments with two drawing rooms at the right of the photograph. UPRR.

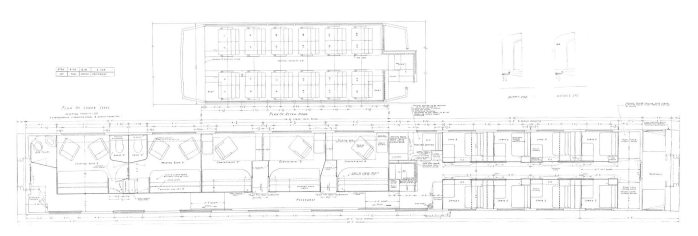

The floor plan of *Dream Cloud*. Pullman Technology, Inc.

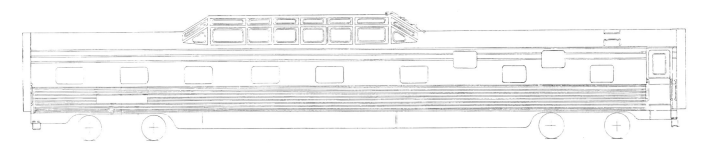

The left side of *Dream Cloud*. Pullman Technology, Inc.

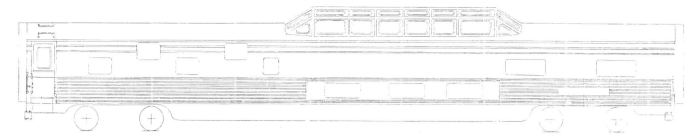

The right side of *Dream Cloud*. Pullman Technology, Inc.

less steel halfway up the wall and the rest of the way is painted green. The ceiling is painted green, too.

The ceiling in the hallway here is painted cream. Three of the walls and the entrance door here at the beginning of the passageway are painted yellow. Now, if you turn around, you can see how the rest of the passageway is decorated down to the staircase at the dome. The lower part of the wall is covered with blue-green leather. There's a stainless steel molding even with the bottom of the window on the outside wall and another stainless steel strip on the inner wall at the same height. From the molding up, the walls are also painted cream. Down at the other end of the car, by the roomettes, the lower part of the wall is painted a blue-green that matches the leather, with the upper wall painted cream. The doors of the roomettes are painted cream, as are the doors to the lockers through the car. The door at the vestibule end is painted blue. They call the color of the carpet Persian Rose. The carpet runs the length of the passageway and covers the first step tread that leads down to the rooms under the dome and the steps to the upper roomettes. The rugs that cover the ramps have a silver pattern. Almost all the lighting is fluorescent.

Let me show you the drawing rooms. The door to drawing room E is painted blue. Each drawing room door is painted a different color.

Drawing room E is my favorite room. There's a big sofa facing the window that folds down into a berth at night. The windows in the drawing rooms are just a little over 5 feet wide. That gives the passengers a big view of the passing sights. Above the sofa is an upper berth that folds down from the wall, and on the outside wall is another upper berth that folds down. The mattresses in all the berths are 35 inches wide. The lower berth mattress is 75 inches long, while the upper berths are 2 inches shorter. With all the berths made up, there's still room for people to sit in the two chairs in the room. They don't have to be folded up and put under the lower berth like you have to do in most rooms. As you can see, the window in this room is a lot bigger than on other railroad cars.

The ceiling and the upper walls are painted beige. Below an aluminum molding, the rest of the walls are painted light blue. The carpet is the same rose color as the carpet in the passageway. The drapes have a pattern they call Chinese Lattice, which has a gray and yellow horizontal stripe with the lattice design at the top and bottom. These are just like the drapes used in the coach sections in the chair car. All the windows in the car have pull-down shades in off-white. The sofa is covered in a blue fabric they call super needlepoint, and the two chairs are covered in a light brown super needlepoint called Sandalwood. One of the nice touches that show these are deluxe rooms is that the armrests and legs on the chairs are made of bleached walnut instead of metal because they don't have to fold up.

Each of the drawing rooms has a lavatory called an annex, which offers privacy instead of passengers having to use facilities in the room like in the compartments and roomettes. Take a look in there. The floor is made of gray-tan Es Es material. The ceiling is light blue, and the walls are beige. The fixtures include a ceramic toilet that dumps onto the tracks through a vacuum valve. The sink folds down for use, and then when you're finished, you fold it up and the water goes down a drain at the back of the fixture and out onto the track. Besides the shaving mirror over the sink, there's a full-length mirror on the door. There is a paper cup holder, and the control panel for the lights, ventilation, and porter's call button is over here by the sofa. In the corner by the entrance door to the room is a full-length wardrobe, and there is suitcase storage above the annex, above the entrance door, and under the lower berth. There's a hat rack over the connecting door to the next room. The traditional shoebox is located at the top of the wardrobe. Since we don't really carry any passengers, I haven't had to shine many shoes on this tour. The only time anyone sleeps on the train is when we move from one city to another overnight. Most of the time we move during the day. The people from General Motors stay in hotels, and the Pullman crew stays on a baggage-dormitory car we call the *Blue Goose*, which travels ahead of the train. When executives from General Motors or Pullman are on-board, I stay on-board to take care of them. There's a porter's section by the staircase to the dome, but I usually use one of the roomettes. Even when we move during the day, I have to be at my post in this car. I'm also on duty during the walk-through tours to answer questions and see that everything stays safe.

There is usually a connecting door with a full-length mirror that goes to drawing room

Drawing room D looking toward the rear of the car. Pullman Technology, Inc.

The annex (lavatory) of drawing room D. UPRR.

D next door, but it has been removed while the train is on tour, so we'll go through here instead of out into the hall. This door allows the two rooms to be made into a two-room suite for large families or friends traveling together, or the door can be closed for privacy. Drawing room D is a mirror image of room E. The only difference is the decoration. In here the ceiling and nosing are painted blue. Then there's a metal molding. Below that, to the floor, the walls are painted gray. The floor is covered with silver carpet. The drapes have little flowers on a white background in a pattern the decorators call Cape Cod Rose. The sofa and one of the chairs are covered with green super needlepoint material. The other chair is covered in something I would call purple, but they call it eggplant. I guess it does have the color of an eggplant. If you open the door to the annex, you'll see that the floor is gray-green Es Es material, the walls are blue, and the ceiling is gray. As you open the door to go back into the passageway, you will see it is painted green to match the room. Now we'll go down this ramp to the compartments under the dome.

There are three compartments below the dome. Just like the drawing rooms, each of the compartments has a sofa mounted facing the window. The window in the compartments is big for a room like this and is 4 feet, 7 inches wide. The first one we come to is

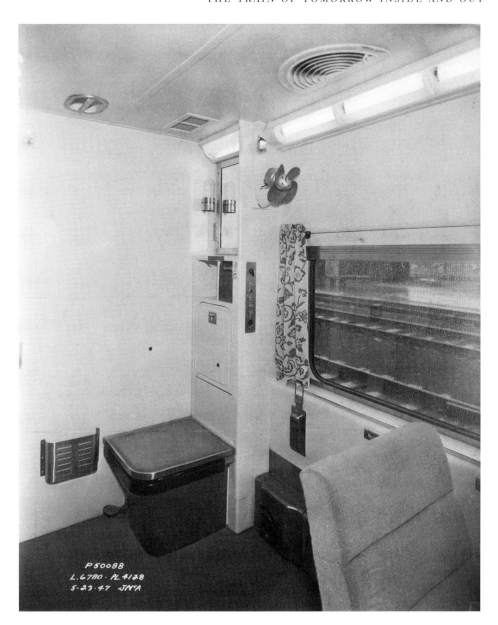

Compartment C looking toward the front of the car. The sanitary column with sink and toilet are next to the window. Pullman Technology, Inc.

room C. Compartments B and C share a common passageway that has green walls and ceiling; the carpet is Persian Rose. The second step down from the main passageway is a glass step lit from below. The glass is three-quarter-inch plate glass that has been sand-blasted on the underside to diffuse the light. The edges and corners are slightly rounded to give the glass a more finished look. The words *Watch Your Step* are printed on it. The door to each compartment is a sliding door, so if you'll slide open the one on the left, we'll go in there first.

This is compartment C. It's equipped with two lower berths. The sofa folds down to make up one of the berths, and the other berth folds down from the wall by the entrance door. The extra chair folds up at night and is stored under the folding bed. In the far corner of the outside wall is what is called a sanitary column. The lid of the little stool lifts up to reveal a stainless steel hopper, or toilet. These hoppers dump out directly onto the track. That is why we ask passengers not to flush while we are in the station. Just like in the annex in the drawing rooms, the sink folds down to be used. The drinking water tap is located below the shaving mirror. On the side of the column is the switch panel for the lights and ventilation. My

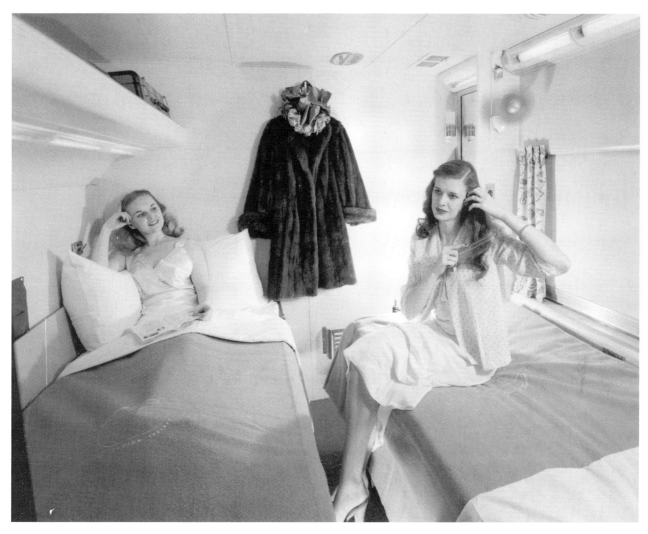

Compartment C made up for the night. The bed on the left is made up from the back of the couch folded down from the wall. The bed on the right folds down from the wall in the foreground at the extreme right. The sanitary facilities are still exposed for use during the night. GM.

call button is also located on that panel. The ceiling, outside wall, and passageway wall are painted ivory. The sanitary column and the two transverse walls are painted cream. The carpet is silver-gray. The pattern on the drapes is called Gladstone and is blue flowers on an off-white background. The sofa is covered in a dark brown needlepoint that is called Chamois. The bolster on the sofa, the chair and the armrests, and the lid of the hopper are covered in a light gold fabric that looks a little bit like corduroy. There is a luggage rack above the sofa, and there is some storage under the sofa even when the chair is folded up under there. If you look back by the door, you'll see the shoe locker box up on the luggage rack. Now let's go across the passageway and take a look at compartment B.

Just like in the drawing rooms, this compartment is a mirror image of the room we just came from. Again, the only difference is the decorations of the room. The color of the carpet is called Dove Taupe. The ceiling and the outside wall to the top of the windows are light blue. From there to the floor and all the other walls are gray-cream except the lower surface of the luggage rack; that is painted red. The pattern of the drapes is called Skywriting and is a bunch of red and black squiggly lines on a white background. The sofa, chair and armrests, and hopper lid are covered in blue needlepoint. The headrest of the sofa is gray needlepoint.

Next to the hopper is a door that connects this room with compartment A. When you open the door all the way, it closes off the passageway so that people can move from one room to the other without going into the passageway. But I can't take you next door to see

A couple in compartment B waits for turndown service while a porter finishes up his work in compartment C. GM.

that room because it is locked off and is being used as the train's office during the tour. When the connecting door is closed, compartment A has a sliding door like the other two compartments. As a matter of fact, I can hear Robert in there on the telephone right now. What I can do is to tell you what the room is supposed to look like when it is put back together for regular service.

Compartment A is laid out just like compartment C. But the decorations are different. The ceiling, outside wall, and passageway wall are blue-gray. The two transverse walls and the sanitary column are painted gray. The carpet is peach, and the drapes have a pattern called Shangri-La that has little white flowers on a pink background. The sofa is covered with a dark turquoise super needlepoint material. The chair and armrests and the cover of the hopper lid are covered with super needlepoint in a soft red called Rose. The sofa has been removed in compartment A and shipped to EMD for storage until the tour is over. In its place there's an office desk, a file cabinet, and a couple of office chairs.

Let's head toward the dome. As we pass the entrance to compartment A, notice that the passageway leading to the compartment off the main passageway has yellow walls and doors. Both the berths in the compartments measure 35 inches by 75 inches.

Watch your step as we go up to the dome. There is one seat up here for every passenger booked on this car and a couple of extras for any guests they might have. During the tour everyone is allowed to sit up here, but in regular service this section would be reserved for the use of the sleeping car passengers only. Since the porter's section is near the stairs, it will be easy to keep track of who's up here. The walls and window frames are painted light blue to go with the blue seat covers. The ends of the seats and the pedestals are painted blue, too. At the front is a dashboard covered in ivory-colored leather. An ivory leather bolster pad runs along the walls just below the windows. There's a clock up there in the dashboard, but I would like to see them put a speedometer up there because I think a lot of people would enjoy knowing how fast we are going. Sometimes it seems like we are just flying along. That diesel locomotive can really move when there are only four cars to pull. The lights in the ceiling down the middle of the aisle are fluorescent. I like coming up here when we are making an overnight movement. I turn out the lights and just watch the passing scenery. I've heard some people call these domes glass penthouses. When you're up here alone at night and the world is passing by, that's just what it feels like. Another interesting experience is being in a thunderstorm up here at night. The lightning goes off all around you, and the sound of the rain on the glass is very restful. Why don't we go back downstairs. I'll show you what I call my office and then the roomettes, and that should just about do it.

Just behind the staircase to the dome is the porter's section, which has a curtain instead of a door. It doesn't afford a lot of privacy, but porters are not usually on a car like I've been on this tour. The curtain is just like the ones used on a lot of Pullman equipment. They call the color Ashes of Rose. On the floor is a jade-colored carpet and the walls and ceiling are painted blue. The seat that is mounted crosswise to the car is covered in brown super needlepoint. There's a low locker next to the seat, and I keep some of my shoe shining supplies in there. Just above that is a large first aid kit that I don't think has ever been used, thank God. Next to the opening into the room is a small locker for my uniform, personal clothes, and some other supplies. On the wall between the locker and the door is the annunciator with chimes and an extension of the intratrain telephone. The annunciator shows me which room has rung for service. Facing the seat is a locker that houses the water cooler refrigeration unit for the cold drinking water that is supplied to all the rooms. The water pump for the system is under the stairs to the dome. Just above that is the bedding for my berth, and above that is a locker for more storage. There is a folding upper berth for the times a porter gets the chance to get some sleep, but I usually stay in one of the roomettes when we are on an overnight trip and there's somebody else on the car. I slept in the porter's section one night when the car was full of people from General Motors and Pullman and we were touring the Southeast.

From here back are the eight duplex roomettes. There are two upper roomettes and two lower roomettes on each side of the aisle. There are two steps up from the aisle to get into an upper roomette. The lower roomettes are at floor level. Let me show you a upper roomette first. This is upper number 8, on the right side of the car if you are facing the front of the car.

This room is equipped with a sofa seat that's just big enough for one person to be comfortable. The seat in the upper roomettes is mounted crosswise of the car and is covered in blue-green super needlepoint. At night a berth lengthwise to the car folds down out of the wall and measures 35 inches by 74 inches. There is a sanitary column with a stainless steel hopper, folding sink, a shaving mirror, a drinking water tap, and a panel of switches for lights, ventilation, and the porter's call button. The lid of the hopper is also covered with green super needlepoint. There's a full-length mirror on the back of the door. When the bed is folded down, the hopper is covered, so anyone wanting to go to the toilet during the night either has to put the bed back up into the wall or use the general use toilet at the forward end of the car. That's why there is a curtain called a Portiere in the hall beside the sliding room door. The curtain is zipped down before the passenger closes the door.

Then, if the passenger wants to raise or lower the berth, they can open the door and step out into the hall behind the curtain. This allows the passenger to stay in their nightclothes to deal with the bed, and then they can close the door again for privacy. It's the same way for the lower roomettes. When the bed is out, the hopper is covered, so the passenger has to do the same thing to use the toilet. The curtain at the door is just like the one at the porter's section and in the same Ashes of Rose color. That color is pretty standard for passageway curtains on Pullman-Standard cars. In the upper roomettes the walls, ceiling, and the inside of the door are painted ivory. The carpet is a tan called Lido Sand.

There are no drapes in the roomettes, but there is an off-white pull-down shade for privacy. There is luggage storage at the top of the wall next to the sanitary column that is actually over the ceiling of a lower roomette. The shoe locker in both the upper and lower roomettes is located between the door and the sofa chair.

Step back here and I'll show you a lower roomette. This is lower number 6. The lower roomettes are smaller because the bed slides out from under the floor of the upper roomette in front of it. It's pulled out just like a drawer. While the overall room design is

The couch of an upper roomette. The bed folds down from the wall behind the couch. Pullman Technology, Inc.

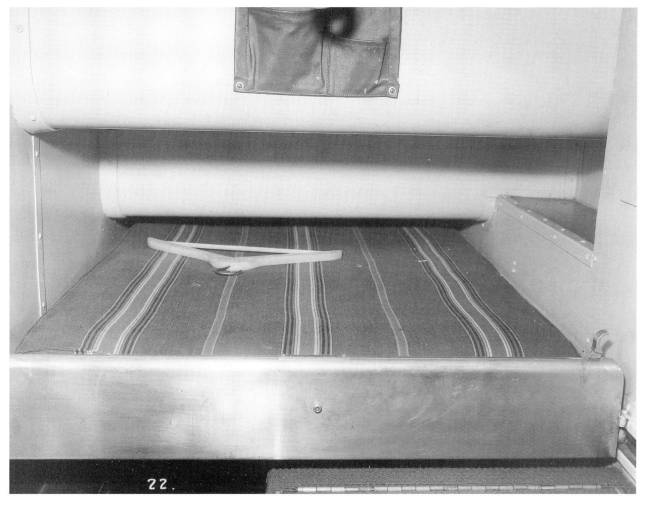

A sliding bed in a lower roomette. When not in use, the bed slides under what is the floor of the upper roomette ahead of the lower roomette. UPRR.

smaller, the length of the berth is only two inches shorter than in an upper roomette. The room is equipped with a sanitary column just like the ones in the upper roomettes. The walls, ceiling, and the inside of the sliding door are painted a light blue-green. The carpet is jade, and the seat and the lid cover of the hopper are covered in brown super needlepoint. The pull-down shade that is used to hide the sliding bed during the day is also brown super needlepoint. There is also a large mirror and light over the bed compartment. They say that with the full-length mirror on the back of the door and that mirror up there on the wall, it makes the room seem bigger. Something about the way mirrors reflect the light. I can't really say I agree that it makes the room seem any bigger, but to some it might.

At the very end of the car are two lockers. On one side of the aisle is a locker for clean linen on the top and soiled linen on the bottom. Across the passageway is a locker that holds clean linen on the top and folding chairs, card tables, and section tables on the bottom. Through the door at the end is the vestibule for this car. The walls are painted gray, and the floor is covered with purple Es Es material with a gray-green inlaid design.

Well, that about does it. Do you have any questions? I think I've covered everything, but I might have missed something you're interested in. If you want to have a seat in the last roomette here, I'll go back to compartment A and see if Robert is ready to take you into the observation lounge car. If not, either I can take you or I'll get someone from in there to do it. It's been my pleasure to show you through here. Like I said before, I'm very proud of this car, and I enjoy showing it to people whenever I can. Most of the time

The sanitary column in a lower roomette with the sink folded down. The toilet is located below the sink under the upholstered lid. At the right is the control panel for the room with switches for the lights, ventilation, fan, a 110-volt receptacle, and a porter call button. UPRR.

people are just passing through on the walk-through tours, and I don't get to spend any real time to show them around. Maybe I'll see you out there in service when this tour is over. I usually do the runs on the *20th Century Limited*. I'll go see if Robert is ready now. I hope you enjoy the rest of your time with us and will come to see us again.

Comfortable? You must be getting tired by now. I don't remember anyone getting such a detailed tour. We usually don't have the time. Usually I do the entire tour, and I know I miss a lot of the things that others could tell you about the areas in which they work. I hope Hill was able to show you everything. If there is anyone who knows about this car, it's definitely Hill.

Well, I got everything straightened out, so I'll be able to finish the tour with you. Let's go take a look at the observation lounge car behind us. As a matter of fact, we'll get something cold to drink in there. I could use a cold glass of soda pop. Watch your step as we go through the vestibule.

CHAPTER 12

MOON GLOW

Well, now we come to the fourth car of the train, the observation lounge car named *Moon Glow*. This car gets a lot of positive comments about how open and airy it seems. The car has enough seats for 68 people. Half the chairs in the car are moveable, so people can make little groupings. When we do the press runs, this is a very popular car. When we started doing the tour, we would open up the bar when the members of the press arrived, and some of them would stop here and not see the rest of the train. Some of them got very drunk, and we would have to take them home. Needless to say, their wives got upset about it. So we waited until everyone had seen the train and then we would open the bar when we got a little closer to our destination. We also made sure we had some kind of snacks for people to munch on so they wouldn't drink so much and to help absorb some of the alcohol. I've seen the time the party got a little rowdy in here, but none of them got out of hand. Smoking is permitted throughout the train, but I have seen it when there was so much smoke in this car you would think there was a fire in here.

Let's start here at the forward end and work our way back to the observation end. I see that some of our invited guests are beginning to arrive, so I may have to excuse myself from time to time to greet the ones I recognize. They're coming for lunch and a tour. Of course their tour will be a lot shorter and a lot less detailed than yours has been. We want to have them fed and out of here before the public tours begin. If you can believe it, there are already about 50 people in line to take the public tour. When I got here at seven o'clock this morning there were a couple of people already waiting. It's been like that at every city we've visited. Sometimes there will even be a large crowd of people at a little town or city that we're just passing through. Pretty amazing, huh?

Let's get started. There are three lounge areas in the car. Here at the forward end is what we call the upper lounge. Under the dome is called the lower lounge, and at the rear end is the observation lounge.

Here, the ceiling is painted a light yellow from the entrance and along the passageway to the rear ramp leading to the observation lounge. The carpet is silver-gray and also runs from the entrance door down the main passageway to the rear ramp. The forward ramp rug has a peach inlay design. The walls are painted gray except for the wall from the entrance door past the ladies' lavatory to the upper lounge. And, as you can see, the inside of the entrance door is painted red.

Knock on the door there and see if anyone is in the men's lavatory. I guess not. Go

The right side of the observation lounge car, *Moon Glow*. In 1956 the Union Pacific Railroad cut off the rounded observation end of the car so that the car could be used in midtrain service next to the dining car. UPRR.

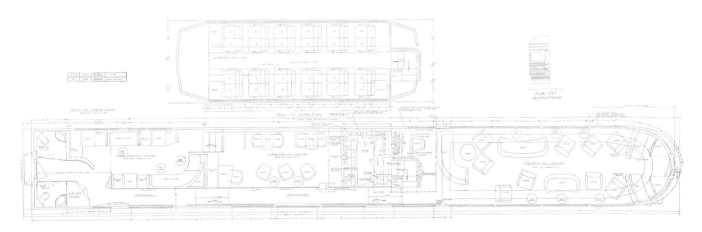

The floor plan of *Moon Glow*. Pullman Technology, Inc.

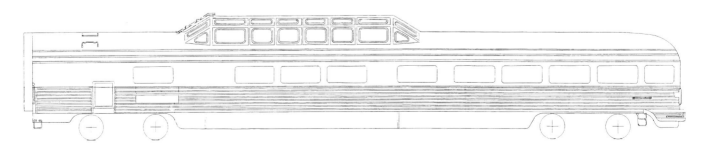

The left side of *Moon Glow*. Pullman Technology, Inc.

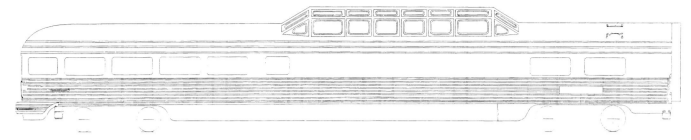

The right side of *Moon Glow*.
Pullman Technology, Inc.

ahead and open the door. The floor is gray-blue Es Es material, while the ceiling is painted light yellow. As in all the general lavatories, the wall is stainless steel from the floor to the wainscot. The inside and end walls are painted rust. The outside wall and the inside transverse wall are covered with a matte-finished V-Board. V-Board is so thin that it can be wrapped around corners and even bent to conform to a curved wall.

I'll knock on this door and see if anyone is in the ladies' lavatory. The ceiling is light yellow and the inside and transverse wall are painted green above the stainless steel wainscoting. The outside wall and end wall are covered in V-Board with a wallpaper pattern with light green apples on it. I think it's funny they call that pattern Adam and Eve. The floor is coral Es Es material.

Let's go around the corner and take a look at the lounges here in the forward end of the car. At the entrance to the upper lounge, in the ceiling of the main passageway, is an edge-lit sign with the words *Top o' the Rail* on it. That's what we call the upper lounge. At the end of the forward passageway is a curved wall that separates the lounge from the passageway, making it feel a little more intimate in the lounge. The side facing the passageway is painted a light yellow, but on the lounge side the wall is covered with something called Weldtex, which is a plywood product with ridges in it to give it some texture, which is then coated to help keep it clean. The pedestals of the built-in seats are also covered with Weldtex. From the transverse partition at the front of the upper lounge, around to the outside wall all the way to the bar on the lower level is covered with a Varlon wallpaper in a Banana Leaf pattern. The banana leaves are both light and dark green on a white background. All the seats in the upper lounge are built-in and consist of couches, booths, and one large curved sofa covered in Chartreuse needlepoint. The bolster and end cushions are covered in red leather. The built-in tabletops of the booths are dark green Formica, and each one has a design in white, black, light green, and tan. One is an idol, one is a dancing girl, and one is a sad monkey. The tables in the lower lounge are the same way. Down there is a table top with a sad monkey, one with a cocktail glass, and another one with a group of ladies playing the drums. For individuals not sitting at one of the booths there are big smoking stands with large tops on them that are big enough to hold a glass or two. The draperies in the upper lounge are made of a material called Satin Mohair.

The ceiling in both lounges is painted ivory, and the carpet is jade green. The inside of the shoulder-height partition that separates both lounge areas from the passageway is painted green, and the finishing rail along the top of the partition is painted ivory. In the lower lounge, the outside wall is painted green from the bottom of the tabletops to the floor. The half-wall partition that separates the two lounge areas and the staircase from the upper lounge to the lower lounge as well as another short partition at the base of the stairs are covered in Weldtex.

The chairs and the built-in sofa in the lower lounge are covered in honey-colored leather. These chairs and the dark green carpet give the lower lounge the appearance of a bar in a private club. A lot of people have complained about the lack of windows in the lower lounge. I have mixed feelings about it. While I think it would be nice to have win-

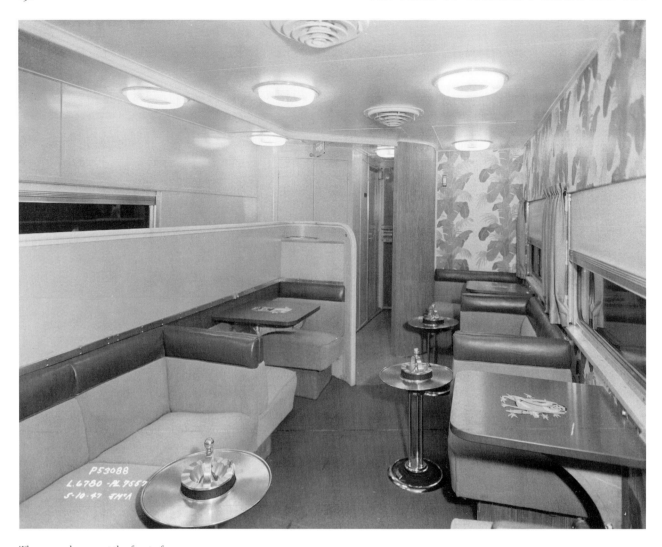

The upper lounge at the front of the car. Pullman Technology, Inc.

dows so people in the lounge could watch the passing scenery, I think the lack of windows adds to the private club feeling.

Now we come to one of the most beautiful pieces of railroad furniture ever built: the bar. While I tell you about it, why don't we go behind the counter and I'll fix us both a glass of cold soda pop. Will a cola do? Good. We have to go out into the main passageway and then back through a door that leads behind the bar. On the wall next to the door we came in is one of the extensions of the intratrain telephone system. Below the door are some of the gauges used to monitor on-board systems.

The whole setup was made by a company named Angelo Colonna. We have a front bar, a back bar, and glassware cabinets above the back bar and to the right side, facing the front of the bar. Under the front bar are two stainless steel sinks, lockers, and a work counter. In the work area on the front bar are bottle holders, porcelain jars to hold garnishes for drinks, a towel ring, a garbage can and chute, bottle refuse cans, an ice cube well, a work light, and a carving board for cutting fruit for garnishes. At the very end of the front bar, on the outside wall, is a plug for an electric iron. It sure has come in handy while we have been out here on tour. I don't know how many times I have needed to press a tie or my coat.

The lower part of the back bar contains the refrigerator and an ice cube maker. The upper part of the back and side bar are the glassware cabinets. On the work counter below the glassware cabinets on the side bar is a cash drawer, a utility drawer, and a humidor for cigars and cigarettes. At the back of the counter of the side bar is an electric plug for a

The upper lounge, looking toward the rear of the car. GM

drink mixer. Up on the wall, next to the glassware cabinet on the side bar is the annunciator with chimes. The facings on all parts of the work area are stainless steel. The counter tops are dark green Formica to match the tabletops in the lounges. I don't know whether you got a good look at the floor in the kitchen and pantry, but this floor is just like that. It is several pieces of metal welded together to form a pan that slopes down to a drain in the center. As a matter of fact, this kind of floor is called metal pan. The metal is also curved up the walls a bit just under the cabinets and lockers to add depth to the pan and make it so there are no cracks and crevices where dirt can collect. With this design, the floor is easy to keep clean. The metal is then covered with Martex, which is an anti-slip material.

Well, all of that is pretty typical for a bar. What makes this so special is the way it is decorated. The glass panels on the front of the bar and on the doors of the glassware cabinets are engraved with images of young ladies in a jungle setting. They add a touch of class and very beautifully carry out the theme of the lounges.

Now we'll go up and take a look at the dome. Turn left as you go out the door. Notice that the color of the rear ramp rug of the passageway changes from gray to peach with a gray inlaid pattern. The carpet is peach all the way to the rear of the car and up in the dome. Turn left here and go up the stairs to the dome. Like the dome compartments in the chair car and sleeper, this has seating available for 24 people. All the painted surfaces are gray. The dashboard and the bolster under the side windows are covered in what is called gray-blue-green leather. I don't know how they can get so many names into one

The lower lounge, looking toward the front of the car. GM

color, but that's what the designers told us in our classes before the tour started. I would call it turquoise, so that shows what I know about interior decorating. They say more men than women are colorblind, but I would think that is taking it a bit far. The seats are covered in turquoise fabric except for the headrests and armrests, which are covered in leather to match the dashboard and bolster. The seats are also equipped with stationary footrests. The ends of the seats and the pedestals are painted gray. The fluorescent lights

The lower lounge featured a bar with wood and etched glass. The photographer was facing the rear of the car, looking down the passageway to the observation lounge. Pullman Technology, Inc.

The dome seating section. GM.

over the aisle can be turned off so that just the aisle lights are on. This allows the passengers to be able to see the passing scenery at night without the glare of the overhead lights.

Now I want to show you what I think is the most beautiful observation lounge on any railroad car. I do a lot of travel by train, and I don't think I have ever seen a room so nicely decorated. When you go down the stairs, turn to the left and let me show you the built-in writing desk area at the base of the dome staircase first.

The writing desk at the front of the observation lounge. The electronic equipment in the cabinet above the desk regulated the intercom, public address system, and music for the entire train. The telephone on the desk was part of the mobile radiotelephone system. Pullman Technology, Inc.

Besides the desk top made of Formica, there are cubicles for writing paper, envelopes, and postcards. We have a clock in the middle of the cubicles, and there are two telephones on the desk, one of which is extension 6 of the intratrain telephone system. The other is a part of the train-to-shore radiotelephone system. As long as we are within a 25-mile range of 30 of the largest metropolitan areas, we can talk to anyone, and I mean anyone, anywhere in the world, that any land-based telephone can call. On the preview run from Chicago to French Lick, Indiana, and back, a call was placed to the ocean liner *Queen Elizabeth*. The captain of the ship and Cyrus Osborn, who runs GM's Electro-Motive Division, wished each other safe trips. As a matter of fact, we were able to broadcast the call over our public address system. The equipment for the radiotelephone is in the cabinet below the desk. A six-volt automobile-type storage battery mounted under the car supplies the power for the radiotelephone system. The antenna is simple: an 18-inch piano wire mounted on the roof. The equipment is supplied to us by Illinois Bell, but it's compatible with the AT&T radiotelephone system around the country.

Now, if you'll step back for a second, I'll open the cabinet doors above the desk to reveal a very interesting surprise. Voila! That's all the electronic equipment we use on the train. The base unit of the setup is the public address system. It has enough power, with the help of some booster amplifiers in each car, to power the speakers on all four cars. A broadcast radio and a wire recorder are linked to the system and provide musical entertainment and news while we're moving. There are two antennas mounted on special supports two feet above the roof of the car. We can make announcements about station stops or comments on the passing scenery by pushing the button on the base of the microphone of the paging system that automatically cuts out the music from the radio or wire recorder. Each speaker can be switched to wire recorder or radio.

There is also landline telephone service available on the train during exhibitions. There is a plug on the outside of this car at the observation end and another on the outside of the sleeper at compartment A, where we have our office while on tour. We have a telephone in the end table on the left side at the observation end and one in the office in compartment A. We carry our own telephone cables just in case there isn't one where we are stopped.

The forward end of the observation lounge. UPRR.

The rear of the observation lounge. Notice how the ceiling conforms to the shape of the car. UPRR.

The whole area is painted gray, including the staircase to the dome, the desk, and the cabinets. The desk chair has a blond wood frame, and the cushions are covered in turquoise super needlepoint. Now if you will turn around, you will see the splendid observation lounge. As you can see, it has an oval appearance. That's as a result of the rounded end of the car and the oval shape of the lighting cove on the ceiling.

There are some features of this room that truly make it one of the most elegant on rails.

Train controls located in the built-in table at the rear of the car. These controls can be used by a conductor or trainman when the train has to be backed up. At the lower left is an air pressure gauge. In the center is the back-up control valve, and at the top right is a handle for a whistle. UPRR.

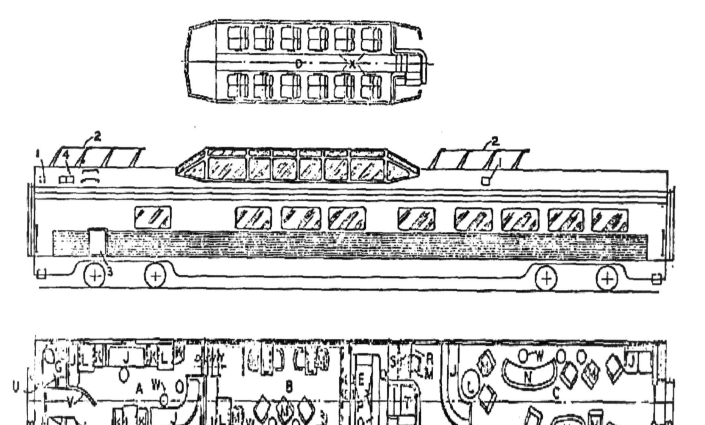

This drawing shows the changes made in *Moon Glow* by the Union Pacific Railroad in 1956. The rounded observation end was cut off so the car could be used in midtrain service as a place for people to wait for a table in the dining car. UPRR.

First, the wainscot, from the heat duct near the floor to just under the windows, is covered in red Velvean Leather with three stainless steel molding strips on it. This goes all the way around the room from the partition here by the passageway to the large built-in sofa just behind the desk and includes the end door at the observation end and the built-in tables on either side of the door. All the painted surfaces in the room are gray except for the oval light cove in the ceiling, which is painted ivory and provides soft indirect and direct light. As I said before, the carpet is peach. The pattern of the drapes has some red, brown, black, green, and gold on a off-white background and is called Persian Tree Chamois.

The nine lounge chairs that can be moved around into groups are covered with gray super needlepoint and have blond wooden frames and legs. There are three sofas in the room. The first is the large built-in curved sofa just behind the writing desk. The other two are small kidney-shaped sofas; one seats three people, and the other seats two. Both can be moved, just like the chairs. When the train is open for public exhibition, we turn the sofas to face into the room so that there is more space for people to move through. But when we are on a run, we turn the sofas and most the chairs around so they face the windows. All the sofas' seat cushions and backrests are covered with turquoise super needlepoint, while the backs and sides of the sofas are covered with light green leather. In front of the built-in sofa is a large round table. The bottom is covered with wood and the big, round top is stainless steel except for the center, which is frosted glass that is lit from below. There are also several big smoking stands in here like the ones in the other lounges.

You'll notice that the windows are large and that there don't appear to be any pillars, but small narrow pillars are indeed hidden behind the drapes. And, of course, all the windows, except the rounded ones at the end of the car, have shades in case the sunlight gets a little too bright.

Let's go back and take a look at what we call the cockpit area at the observation end of the car. This door is used as the entrance for the public tours. A ramp, like the one at the locomotive, is put up outside and a large padded bumper is put at the top of the door frame in case anyone bumps his head. On either side of the door is a built-in end table, and facing those are built-in seats that are bowed to match the curve of the tables. Like the sofas, these seats are covered with turquoise super needlepoint fabric. The backs, which have magazine pockets in them, and the sides are covered in the same light green leather as on the couches. Let me open this door on the aisle on the table on the right and you'll see a back-up and signal valve that allows the conductor or one of the trainmen to come back here and help control the train when we are backing down a track. That little entry door facing the seat allows access to a steam valve shutoff handle.

Well, I believe that does it. That's the *Train Of Tomorrow*. Do you have any questions? No? Then how about lunch? We have about 90 minutes before the public will start moving through the train. Buddy Tyson told me he has a couple of steaks for us, and I know Mr. Cruickshank has a good bottle of burgundy set aside as well. We won't have to suffer through Chicken Cordon Bleu. I understand there is one empty table in the dome in the dining car waiting for us. So unless you can think of anything else you would like to see, why don't we head for a delicious meal in the sky?

APPENDIX A

THE DIARY

While the *Train Of Tomorrow* was on tour, someone kept a daily log, listing such things as movements of the train, notes about stations and show locations, special guests and unusual activities, as well as the number of people to take a tour of the train. The diary covers the entire 28-month tour throughout the United States and Canada, from the preview run to French Lick, Indiana, to the final trip back to Chicago where the train was put up for sale.

It is not known who kept the log for the *Train Of Tomorrow*'s tour, but the writing style is similar throughout, so it can be surmised that only one or two men were involved. All errors, typos, and grammatical mistakes have been preserved.

Abbreviations used in appendix A

ACL: Atlantic Coast Line Railroad
B&A: Bangor and Aroostook Railroad
B&M: Boston and Maine Railroad
B&O: Baltimore and Ohio Railroad
BOP: Buick, Oldsmobile and Pontiac Division of General Motors
C&EI: Chicago and Eastern Illinois Railroad
C&NW: Chicago & North Western Railway
C&O: Chesapeake & Ohio Railroad
C&WC: Charleston & Western Carolina
CB&Q: Chicago, Burlington & Quincy Railroad
CNR: Canadian National Railroad
CPR: Canadian Pacific Railroad
CRI&P: Chicago, Rock Island and Pacific Railroad
D&C: Detroit and Cleveland Navigation Company
D&RGW: Denver and Rio Grande Western Railroad
DL&W: Delaware, Lackawanna and Western Railroad
FEC: Florida East Coast Railroad
Frisco: St. Louis and San Francisco Railroad
GT: Grand Trunk and Western Railroad
IC: Illinois Central Railroad

Katy: Missouri-Kansas-Texas Railroad
L&N: Louisville and Nashville Railroad
NC&STL: Nashville, Chattanooga and St. Louis Railroad
NYC: New York Central Railroad
NYNH&H: New York, New Haven and Hartford Railroad
NYO&W: New York, Ontario and Western Railway
P&LE: Pittsburgh and Lake Erie Railroad
PRR, Penna: Pennsylvania Railroad
RF&P: Richmond, Fredericksburg and Potomac Railroad
SAL: Seaboard Air Line
SP: Southern Pacific Railroad
T&NO: Texas and New Orleans Railroad
T&P: Texas and Pacific Railroad
TH&B: Toronto, Hamilton & Buffalo Railway
WP: Western Pacific Railroad

Preview Run–May 26, 1947
We left Dearborn station in Chicago on the Monon R.R. at exactly 10:30 AM. The weather was clear and bright. Shortly after departing, a passenger asked to be permitted to jump from the train as he had only come aboard to inspect the train. Result, we stopped at Englewood.

Inspection of the train was in progress when the first luncheon call was received. During the trip all were invited to ride in the cab of the locomotive in groups of two. The streets of each town were lined with people to view the train as we passed through. Near Greencastle, Indiana, our engine crew noticed some tie plates piled on the track but too late to stop. The locomotive took them in stride and passengers were unaware of the incident. Due to several courtesy stops we arrived in French Lick about thirty minutes late. 758 hotel guests inspected the train in the evening.

May 27, 1947
Lost baggage caused a ten-minute delay in our start from French Lick on our homeward trip to Chicago. Stop made at Bloomington for water. At a point near Lafayette a chain extending down from a water spout struck the steel sides of the dome framework on all cars but no damage resulted. Stop was made to allow photographers to take a "profile shot" of train. At about 2:30 PM slight automobile accident occurred but no one hurt. This may prove to be our full quota of accidents. As we were approaching Dyer, Indiana, Mr. Osborn made his radio telephone connection with friends on the *Queen Elizabeth*, then 2,400 miles at sea. Recording made. We arrived at Illinois Central Station 6:15 PM.

May 28–June 1, 1947
Wednesday, 28th Christening ceremonies and invited guests' inspection. Mr. C. F. Kettering officiating.

Thursday thru Sunday—public exhibition. 50,000 enthusiastic people.

June 2, 1947
Train cleaned and readied for start of eastern tour.

June 3, 1947—1st Eastern Tour
Left Dearborn Station, Chicago at 9:15 AM on Grand Trunk and Western R.R. for Detroit. Many prominent railroad officials aboard. Mr. Budd, president of the Burlington, talked to his son in Detroit via Radio Phone to advise of his arrival time. Mr. Walters of the Canadian National talked to his office in Montreal while we were about fifty miles out of Chicago.

We exhibited to 260 people in the short length of our stop at Battle Creek. At Lansing, our motion picture group left us to return to Chicago. Here we were joined by approximately 160 members of the press, radio and GM executives who rode with us to Detroit. Short stop outside of Lansing to allow a front view shot of moving train. About ten minutes out of Durand we took a broadside from a small boy's rock. Damage consisted of one outside sheet of side dome glass. Arrived at Detroit 5:45 PM.

June 4–June 8, 1947
Train exhibited to 60,000 people in spite of much rain. Coal dust adds to crowd's discomfort. Mr. L. E. Cowles lives up to his reputation for seeing to details and nothing has been overlooked.

June 9, 1947
We left Detroit at 8:45 AM for Toledo. About 180 guests aboard consisting of railroad executives, GM officials and some news and radio men and women including Mr. Gudrun Hedgaard, news correspondent from Norway.

After a short stop at Monroe and two slow orders we arrived in Toledo at ten. We were met by officials of the Libby-Owens-Ford Glass Co., news correspondents and officials of the B&O R.R. Ramps set up at the B&O team tracks at the foot of Washington St. and opened for the luncheon guests at 2:30 PM. We had 11,503 guests and closed the line at 9:00 PM. At 10:00 PM about 500 people had been turned away.

June 10, 1947
Left Toledo at 1:00 AM and arrived at Akron 5:00 AM. Opened to the public at 9:00 AM and had 2,764 visitors in three hours. Left for Youngstown at 12:00 PM and arrived at 1:30 where we were visited by 785 people before the press came aboard. Mr. Osborn, who flew from Chicago by private plane, joined us here. Departed at 3:10 with about 100 members of the Pittsburgh press and radio as well as railroad officials. Stopped at New Castle and Elwood City and arrived at Pittsburgh at 6:00 PM. Four radio broadcasts were made enroute with Mr. Osborn, Rukgaber, Fitzgerald, and Hamilton. Several people were interviewed including Mr. Daniels and Mr. Perkins.

June 11–15, 1947
Exhibited to 39,258 people. Some rain but excellent location kept discomfort at a minimum.

June 16, 1947
We were gliding out of Pittsburgh at 7:00 AM on a beautiful sunny morning. Breakfast served for all aboard by the Pullman masters of the craft. The B&O route followed was a real example of the beauty to be seen from the Astra Domes.

Stopping at Cumberland for one and one half hours, 2,155 people came aboard. Unable, because of the tight schedule, to accommodate the additional 500 who remained in line. At Martinsburg we were joined by 135 members of the Washington press and radio staffs. Broadcast made enroute to Silver Spring. Arrival time 4:30 PM.

June 17, 1947
Luncheon aboard "Sky View." Guests: Railroad Executives. Host: Mr. R. K. Evans.

June 18, 1947
Luncheon aboard "Sky View." Guests: Government Executive Staff. Host: Mr. C. E. Wilson.

June 19, 1947
Luncheon aboard "Sky View." Guests: Congress. Host: Mr. Paul Garrett.
Breakfast aboard "Sky View." Guests: President's Committee for Highway Safety. Host: Mr. Frederick C. Horner.

June 20, 1947
Luncheon aboard "Sky View." Guests: U.S. Army, U.S. Navy. Hosts: Mr. C. L. McCuen, Mr. G. W. Codrington.

June 21, 1947
Luncheon aboard "Sky View." Guests: Foreign diplomats. Host: Mr. E. S. Hoglund.

June 22, 1947
Left Silver Spring Station at 12:01 AM proceeding to Atlantic City via Baltimore and Philadelphia. Crew changes at Gray's Ferry (entering Philadelphia and at Frankford Jct. leaving Philadelphia). Riding in Astra Dome prohibited because of low overhead electric lines of PRR through Philadelphia. Daylight breaks as we crossed the Delaware and start a non-stop run across the Jersey marshes to Atlantic City. Arrive at 5:00 AM PR&SL Station.

June 23–27, 1947
Participate (and steal the show) in the Railroad Suppliers' Manufacturers Association convention track exhibit located Bachrock and Ohio Avenue. Extremely dirty from coal smoke and cinders.

June 28, 1947
Left Atlantic City at 12:01 AM with only our own crew aboard. Proceed to Wilmington Plant of BOP Assembly Division via Philadelphia and Gray's Ferry. Arrival time 3:00 AM. Exhibit to GM employees 10:00 AM–9:00 PM.

June 29, 1947
Public exhibit to 12,816 people.

June 30, 1947
Train serviced and cleaned completely. Mr. Lou Cluxton of Frigidaire presents DuPont Hotel with black cat.

July 1, 1947
Leave Wilmington BOP Plant at 1:55 PM with luncheon guests including Mr. Sloan's party from New York. Stop at Wilmington B&O Station to pick up press and radio people from Baltimore. Broadcast made enroute. Nearly all guests visited cab of locomotive for a short ride there. Arrive at Camden Station in Baltimore at 4:02 PM.

July 2–7, 1947
Exhibit at Camden Station of B&O R.R. Roy Westin votes this our best stop so far.

July 8, 1947
Leave Camden Station at 2:30 AM and after an hour we find ourselves delayed in Potomac yards awaiting green light to proceed into Virginia. Arrive Fredericksburg at 6:00 AM and make ready for 9:00 o'clock exhibition. Wooden ramps erected on the sight. We are joined at 12:00 noon by Richmond press and radio party and proceed to Richmond via RF&P route. Arrive Richmond 3:00 PM. Exhibition time 6:00 PM with our own ramp equipment.

July 9, 1947
Exhibited at Richmond 12:00 to 9:00 PM.

July 10, 1947
Leave Richmond 12:30 AM for long hop to Charleston via ACL. Step up tickets are purchased for crew, necessary for night travel when sleeping car is in use. Arrive Charleston 9:30 AM. Hotel Ft. Sumter for crew members at this stop is greatly admired. Luncheon and exhibition 2:00 PM to 9:00 PM. Real crowd, two lines of people, our first encounter with the Deep South.

July 11, 1947
Leave Charleston 9:11 AM after short wait for crew members, detained by freight train across street near depot. Press board us at Yamassee, Georgia, for ride into Savannah, arriving at 11:30 AM. Luncheon, Mr. J. E. Johnson, speaker, 12:15 and public exhibition from 2:00 PM to 9:00 PM. ACL personnel act as hosts and hostesses.

July 12, 1947
Crew members enjoy DeSoto Hotel swimming pool in the morning and we leave Savannah at 12:01 PM with members of Jacksonville press and radio aboard as well as several ACL officials. Arrive at Jacksonville 3:00 PM. Very warm. Exhibit at Jacksonville Terminal beginning 6:00 PM.

July 13, 1957
Exhibit Jacksonville 2:00 PM–9:00 PM. Mr. Kunkle and Mr. Hamilton arrive early evening via GM plane.

July 14, 1947
Train cleaned and serviced. Luncheon guests, with Mr. Kunkle as host inspect train from 2:00 PM until 4:30 PM.

July 15, 1947
Leave Jacksonville at 8:10 AM for all day run to Atlanta. Train boarded by ACL officials and party at Waycross. Tremendous ovation here. Literature distributed among the crowd at the depot and generously scattered from train to crowds at way stations not due for scheduled stop. ACL has generated public interest greatly. At Fitzgerald, Ga., we are the recipients of twenty huge watermelons, the gesture of Chevrolet people in town. At Manchester it was two bushels of the famous Georgia peaches. Here, too, we pick up Atlanta press folks at 3:00 PM. Rain has not to any degree diminished the crowds at various depots. Arrive Atlanta 5:00 PM. Henry Grady Hotel for the crew.

July 16–20, 1947
Public exhibition at Union Station.

July 21, 1947
Leave Atlanta on L&N tracks at 12:01 AM for night run to Knoxville through the edge of the Great Smokies. Arrive 5:00 AM for one day exhibit. Open 12:00 noon for GM and railroad people. Public showing 2:00 PM–9:00 PM. Andrew Johnson Hotel is rich in historic lore but very short of vitamins. Mr. H. Dunsford confined to Roomette #5 for the next day.

July 22, 1947
Knoxville begins to recede in the distance at 9:00 AM when we glide from Union Station with a small number of L&N officials aboard. Louisville press board us at Lebanon, Ky., for sixty-mile ride in Louisville, arriving 6:00 PM.

July 23–27, 1947
Exhibit at L&N Station, Louisville. Brown Hotel is most hospitable.

July 28, 1947
The L&N Railroad are again our hosts on our swing up the Ohio to Cincinnati. Leave Louisville at 8:10 AM and are soon "hitting it up" as we wind in and out among the hills of Kentucky. A number of L&N officials accompany us on this hop and are duly impressed with the quality of ride over this turning, twisting stretch of track. Arrive at Cincinnati at 11:15 and are spotted on track #2 at the awe-inspiring Union Station.

July 29–31, 1947
Unusually good attendance for these three days in almost tropical heat. The 4-H Club in boarding their train almost got our crowd out of hand. They were allowed to cross our line and in doing so our crowd broke away and attempted to follow. Ten minutes of confusion.

August 1, 1947
Although our schedule called for a ten o'clock departure for Dayton, one of our crew, Mr. A. A. Daniels, Service Manager, failed to heed the alarm and as a consequence saw the Train Of Tomorrow as we drew out of Union Station, but from the platform a block or

more in our wake. Said gentleman now has various phone calls, reminding him of departure time on moving days.

The trip to Dayton, while short, was a pleasure to our crew, in that we welcomed back many old friends who had at some time been our guests or active members of this activity. Arrival time at Dayton was 11:30 AM and we were placed on the B&O team track at the lower level of the Union Station. A wonderful location and every detail planned by Mr. B. J. Van Doren as to make this one of our most successful exhibitions. This was especially gratifying to the crew, for it was here that we said goodbye to Mr. Geo. Rukgaber, our Manager, who left us to take a well deserved holiday from the train. The Pennsylvania hills must have been echoing for him with Cy Perkins' blasts of the horn on our #765.

August 2–5, 1947
GM employes 10:00 AM–1:00 PM.
   Public 2:00 PM–9:00 PM. Total showing approx. 55,000.

August 6, 1947
Farewell to Dayton and on west to Indianapolis for the last showing of our southeastern tour. Back track on the B&O to Hamilton and then on via Connersville where we are joined by the press and radio group, 100 strong. Arrive slightly late at 5:15 PM because of an unscheduled stop at Oxford, O.

August 7–10, 1947
Exhibit at Fair Grounds. Weather continues to be very warm. Mr. Al Meyers is "officer of the day" in the absence of Mr. George Rukgaber.

August 11, 1947
At 12:01 AM we are again pounding the rails of the Monon for Chicago and the Pullman Shops. Aboard, in addition to the crew, are a few Indianapolis citizens who take this opportunity to enjoy a beautiful ride under the stars in our Astra Domes. Upon arrival at Hammond our Train Of Tomorrow is uncoupled from the locomotive which goes to the shops at LaGrange and we are drawn by "foreign power" to the Pullman yards. Arrival time, 6:00 AM.

The first leg of our nationwide tour is over. All has gone well and thanks to a very receptive public, the Train Of Tomorrow is fast becoming an American byword.

August 11–19, 1947
T.O.T. in Calumet shops of the Pullman Co. for complete overhauling and various minor engineering changes as indicated necessary by our road experience on the southeastern tour.

August 20, 1947 — 2nd Eastern Tour
The T.O.T. stands at Grand Crossing ready for the second leg of the tour. As a handful of Public Relations personnel, crew members, and NYC officials await the approach of the train at 63rd Street Station of the New York Central at Chicago, there is much small talk as to the reason for the delay in departure. The T.O.T. approaches the station at 7:35 AM and we are ready for the second trip east. The 35-minute wait is caused by failure to replenish the water supply.

Leave 63rd St. Station at 7:40 AM and on the excellent tracks of famous 20th Century Limited the T.O.T. seems eager to run at unlimited speed. A short stop in Elkhart, Ind., to allow four courtesy passengers to detrain and we are rolling towards Toledo. Arrive Toledo at 11:30 AM and lay in siding to await arrival of Cleveland press group due at 1:35 PM. Group on time and at 1:45 PM we are again rolling eastward with 95 members of press fraternity aboard. Various slow downs made thru towns to allow local populace to view train as it glides by. Arrive Cleveland 4:45 PM old station near waterfront.

August 21–25, 1947
Exhibit at D&C dock tracks near Stadium. Train has to be moved out every night to allow coal train to enter docks.

August 26, 1947
At 11:00 AM we slowly draw away from site and wend our way through the maze of tracks toward Buffalo. Lunch served aboard, for train staff and a few NYC officials, in the Astra Dome of Sky View.

Arrive at Erie 2:00 PM and are soon boarded by Buffalo press group. Leave Erie at 2:40 PM. Much interest shown by townspeople as we roll through this beautiful section of Western New York State. Arrive Buffalo at 4:30 PM.

August 27–31, 1947
Exhibit on tracks just off NYC station site near Blvd. Chevrolet does a fine job of cleaning up grounds.

September 1, 1947
Early morning departure, 7:15 for Toronto via Welland on the NYC. Then to Hamilton on the TH&B and from there to Toronto on the Canadian Pacific road. For the first time the T.O.T. is to be on foreign rails. Difficulties experienced here as we cross the border, concerning the entrance of our Philipino chef, Wm. Schwartz, and resulting in his labor day turning into a holiday and he returns to Chicago to join us again when we return to U.S. We proceed to Toronto, arriving at 11:00 AM. Luncheon served aboard train to 80 Canadian dignitaries. Kitchen presided over by our very able Inspector of Cuisine, Mr. Joseph Schneiderbauer.

September 2–6, 1947
Exhibit at Toronto Exposition on special tracks just inside the Dufferin Gate. T.O.T. is most popular attraction among a host of marvelous exhibits. Popular expression of opinion is "Smart train, eh what!"

September 7, 1947
Bid farewell to Toronto at 9:00 AM and retrace our "steps" to Hamilton and then to St. Catharines for a single day slowing sponsored by McKinnon Industries. Arrive 11:20 AM. Very impatient crowd makes this a difficult showing.

September 8, 1947
Leave St. Catharines at 4:00 AM and roll slowly toward Suspension Bridge and back into United States and are again on NYC tracks. Short wait for signals at the bridge and at 8:30 we are on our way to Lockport to pick up our press group. Arrive 9:00 AM and leave quickly for Rochester pulling up at Read Blvd. to detrain our guests at 10:30 AM. Train is turned and we back into the site at Ontario Beach Park at 12:00 noon.

September 9–11, 1947
Exhibit on Municipal Warehouse tracks beside Canadian Steamship docks at Ontario Beach Park. Tracks have been extended 100 ft. to accommodate T.O.T. by courtesy of New York Central. Wonderful crowds, notwithstanding our location nine miles from heart of Rochester. Train Mgr. Rukgaber called to Montana by illness of his brother. Mr. G. M. Coffin of Detroit makes us a short visit.

September 12, 1947
Leaving our site at 8:20 AM we roll to Rochester Station where a few invited guests board for the ride to Syracuse. The city fades in the distance as we leave at 8:50. Stop made at Newark at 9:30 to pick up press group of 60 persons. Arrive in Syracuse at 10:30 and are spotted on Perishable yard tracks new Brown-Lipe-Chapin Plant.

September 13–15, 1947
Exhibit Friday through Sunday under direction of Mr. W. E. Hamilton in the absence of Mr. Rukgaber.

September 16, 1947
T.O.T. still pointed eastward down the beautiful Mohawk Valley as we leave Syracuse at 12:50 PM. Press pick up at Utica 3:30 and continue on down the valley to Albany. GM photographers busily at work on this run. Arrive Albany 5:15 as darkness gathers. Train spotted at foot of Columbia Avenue, north of NYC Station on the west bank of the Hudson. Mr. L. H. Kurtz among guests.

September 17–21, 1947
Exhibit Tuesday thru Sunday. Governor Dewey and party on 18th.

September 22, 1947
At 8:00 AM we are backing out of Albany. Train is turned on NYC bridge and we return to Rotterdam Jct. where we switch to Boston and Maine rails for our run into Boston. It's a dismal wet Monday morning and after much delay we are off for Boston at 10:50 AM. Our route takes us through the Berkshires and the long Hoosick tunnel. For just a few miles and fewer minutes we are crossing a corner of Vermont. Welcome 69 press people aboard at Gardner, Mass., at 3:30 PM. Draw up at North Station, Boston at 5:10 PM.

September 23–28, 1947
Exhibit at North Station. September 27th we move up to Lowell Jct. to allow photographers some moving shots of T.O.T. Forty miles round trip. No passengers aboard. Lock the doors, Mr. Elmer Krueger has lost his hat.

September 29, 1947
Staff boards train at midnight and at 3:00 AM we are swiftly rolling toward Worcester, where we leave B&M tracks for the Boston and Albany road at 5:00 AM. Breakfast served aboard at 8:30 while twisting about the hills of western Mass. After a service stop at Springfield, arrive Chatham, N.Y. 11:00 AM and leave at 12:30 PM on N.Y.C New York press pick up at Middleton made at 2:30 PM. Follow Harlem Division to Putnam Jct. and Putnam Division into Van Cortland Park Station arriving at 5:15 PM. Guests detrain here and T.O.T. goes to 60th Street yards for fueling and is "Y'd" near the Spit and Divil bridge. Leave New York to return to Putnam Jct. where we have an overnight layover. Arrive Putnam 10:50 PM.

September 30, 1947
T.O.T. is met by special train at 9:15 AM carrying second press group from New York. Leave Putnam Jct. at 9:30 and arrive Van Cortland at 11:30. Train again is turned and fueled, returns to Van Cortland at 12:45 for pick up of Mr. Sloan and invited guests. T.O.T. quickly departs for Putnam Jct. to be turned and again make the run into New York with honored guests. Luncheon is served as party boards the train. Arrive Van Cortland Station at 4:45 PM where guests detrain and T.O.T. proceeds to site of exhibition at 61st Street and West End Ave. via tunnel beneath Riverside Drive.

October 1–6, 1947
Exhibit at 61st St. and West End Ave. Slow starting crowd grows as days pass and near capacity is reached on final days showing. World Series games are tough competition and contribute to our very unsatisfactory hotel accommodations the first two days of our stay in New York.

October 7, 1947
At 12:01 AM we are under way on our jaunt up the east bank of the Hudson to the bridge at Selkirk where we cross at 6:30 AM. Proceed slowly down the west side of the river toward Weehawken, only 10 miles from the site of our New York show. Elizabeth group board us

here for the ride to Elizabeth, N.J. Arrive at the B&O freight yards at 12:40 PM. Showing to GM employees and railroad people from 3 to 9 PM.

On this trip we are accompanied by four men from Sound Masters, Inc., who are busily engaged in picture taking. Weather ideal. As we roll down the Hudson River Valley past West Point and other famous scenes we are paced by Sound Masters small plane just some 50 feet above the river. Cameramen working hard to get just the right shots. Some very good color film should result.

October 8, 1947
Leave Elizabeth site at 12:40 PM, backing out to main line of B&O. Stop made at station for press group from Elizabeth plus numerous GM people. This group ride with us to Wayne Jct. Philadelphia where they unload at 2:45 T.O.T. continues on to 24th St. Station of B&O. Here we are hosts to Philadelphia press group on round trip ride to Elsmere Jct. and after detraining at the station 5:00 PM the T.O.T. proceeds to site of exhibition at South Philadelphia.

October 9–13, 1947
Exhibit at South Broad St. across from Stadium, near the Navy Yard. Location not satisfactory from standpoint of cleanliness, accessibility, nor background. This is the last of the exhibitions in the Region of Jim Bird. Good newspaper and radio coverage has been a feature of the shows in his territory. Mr. Herb Grenda, new assistant to Mr. Bird, makes his debut with the T.O.T. at this Philadelphia show. Nice going, Herb.

October 14, 1947
Back out of site at 8:30 AM to the B&O main line and proceed toward Baltimore. Four guests and train staff aboard. Pass site of our Baltimore exhibition and also our show at Washington (Silver Spring, Md.). From the Astra Dome can be seen the Capitol Dome and Washington Monument as we glide thru the Nation's Capitol. Short stop made at Cumberland at 2:00 PM for watering air-conditioning condensers. Leave Cumberland at 2:20 for Rockwood arriving 3:40 PM. Here we leave the B&O main line for Johnston, Pa. Poor track and slow ride. Train is turned at a point 15 miles from Johnstown and we back in, through the city and into the yards of the Carnegie-Illinois Steel Corporation at 6:15 PM. Much interest shown by townspeople as we enter the town. While we are backing our way into Johnstown and are emerging from the last curving tunnel it is said that the tunnel shifted just a fraction of an inch and we roll back a chrome strip on the side of Sky View. Track is moved by workmen so this will not occur when we leave via the same route next morning.

October 15, 1947
Leave Carnegie yards at 8:20 AM with Carnegie executives and wives as well as several GM executives. Breakfast is being prepared aboard as the Johnstown High School band bids us farewell. We are on our way down the rocky road to Rockland Jct. and one again on "high iron" rolling to Pittsburgh and the Homestead plant of Carnegie-Illinois for the annual open house. Arrive at 1:15 PM on P&LE tracks since leaving McKeesport, 10 miles back.

October 16, 1947
Exhibit at Carnegie Steel, 14,625 people file through the train for one of our most successful days. Leave Homestead at 10:30 PM and are bound for Chicago over the B&O.

October 17, 1947
Arrive Pullman Jct. at 6:45 AM. Locomotive goes to LaGrange and train to Plant #2 by switcher. Arrive Plant #2 10:50 AM.

October 18–23, 1947
Train overhauled and refurbished at Plant #2 by Pullman craftsmen.

October 24, 1947
On display at Electro-Motive Division open house at LaGrange Plant.

October 25, 1947
Leave LaGrange at 10:45 PM (24th) and proceed to Detroit via Wabash route. Arrive Detroit 8:00 AM and after cleaning in the 21st yards we back into Union Station to receive Mr. Wilson's guests, the Associated Press Editors, for their trip to Ann Arbor to attend the Minnesota-Michigan Football classic. Leave Detroit 12:21 on Wabash via Milan. Switch at Milan to Ann Arbor R.R. and thence to stadium. Guests unload at 2:10 PM and train is backed to Milan, turned and backed into stadium tracks for return trip. Guests board train after game as they are saluted by the U of M band. Staff members were graciously supplied with tickets for the game. Departure time 4:55. Arrival 6:10 PM Union Station. Leave Detroit 8:45 PM, arrive back at LaGrange 4:45 AM (26th).

October 26, 1947
Exhibit LaGrange 10:00 AM–10:00 PM.

October 27, 1947—Western Tour
Train leaves LaGrange at 1:30 AM arriving 14th Street yards 4:30 AM remaining until 8:20 when we pull into Burlington Station at 8:30. Leave Chicago 9:30 AM with Mr. C. R. Osborn hosting railroad executives. This group of officials detrain at Galesburg 11:30 AM. We have made up the ½ hour lost in the late departure from Chicago by means of 90 mile an hour speed over the great Burlington roadbed. Leave Galesburg at 11:40 with a few press people aboard for the ride to Burlington, Ia. This being one of the few rainy days of travel we are not accustomed to seeing the T.O.T. with a "dirty face." Short stop at Burlington 12:40 PM and another at Ottumwa 2:10 PM. Press from Omaha and Council Bluffs board train at Creston 3:30 PM and we are rolling into the station at Omaha at 5:40 PM.

October 27–30, 1947
Exhibit on track #9 at Burlington station, Omaha. The gateway to the west. Wide streets, good food, and even better people.

October 31, 1947
The Train Of Tomorrow left Omaha at 9:15 AM for Lincoln. Mr. Daniels at 9:30 via Greyhound for the second time. Guests include Governor and Mrs. Val Peterson. Arrive Lincoln at 10:30 AM. Electro-Motive Division is host at 12:15 luncheon. Mr. N. C. Dezendorf, speaker.

November 1, 1947
Public showing at Burlington Station, Lincoln. Nice radio plug announcement made at Nebraska-Missouri game.

November 2, 1947
Leave Lincoln at 10:30 AM for Akron, Colorado. Stop made at Hastings for 3 hour showing at 12:30 noon. Capacity crowd. Leave Hastings at 5:00 PM arriving at Akron 8:45 PM for all night lay over. Night life in Akron at a minimum.

November 3, 1947
After fueling and watering the leaving time is 9:30 AM. Fort Morgan stop made at 10:00 for Denver press pick up. Arrive Denver 11:45 AM. Light snow, our first.

November 4–6, 1947
Exhibit at Wazee Market tracks of D&RG.

November 7, 1947
Special picture run made to Colorado Springs in forenoon. Approx. 25 models together with a Pikes Peak background made possible some fine color shots, both inside and out.

Return to Denver at 1:00 PM for special cleaning. Two broken springs discovered, ordered from Chicago and are replaced by midnight.

November 8, 1947
Leave Denver at 7:15 AM (15 minutes late) and arrive Colorado Springs on schedule. Railroad officials, guests of the trip numbered 70. Train staff took the opportunity to visit Will Rogers Memorial, Garden of the Gods and other local points of interest. Both the city of Colorado Springs and the surrounding country proved a camera fans' paradise.

November 9, 1947
Left Colorado Springs at 7:00 AM for the short run to Pueblo. Exhibit 10:00 AM–10:00 PM. Much of the state's hard labor performed in Pueblo.

November 10, 1947
Mr. Osborn, Mr. Earl and Mr. Evans with their wives are aboard with special party from Detroit, and Chicago, as we leave Pueblo at 8:00 AM for the much anticipated trip through the Royal Gorge of the Colorado. Weather somewhat short of perfect but the sun shown at times, making picture taking a nerve wracking business. Many guests of the Denver and Rio Grande aboard for short stretches of the trip. This is truly Astra Dome terrain. Arrive Grand Junction in late afternoon for the three-hour showing before facing the rigors of the local hotel. At this showing the D&RG provided an excellent set of wooden ramps. Fit perfectly.

November 11, 1947
Left Grand Junction at 9:15 AM for Salt Lake City. Stop at Thistle 2:45 PM for press pick up and continue on through a driving snow. Arrival time in Salt Lake City 4:45 PM.

November 12–14, 1947
Exhibit at D&RG station. Luncheon of November 12th—Mr. J. E. Johnson, speaker.

November 15, 1947
Left Salt Lake City at 10:00 AM. Arrive Ogden 11:30 AM. Because of a wet snow our guests at the showing entered at the locomotive end of the train and the observation car used as exit point. The line which formed on the sidewalk near the front end of train saved both the crowd and the train carpets from the muddy approach near the regular entrance ramp at rear end of the train.

November 16, 1947
Leave Ogden at 12:30 AM, retracing to Salt Lake City via D&RG tracks. Here we are delivered to the Western Pacific R.R. for the remaining distance to Oakland, California. Train staff sleeps aboard and enjoys breakfast in the Astra Dome at 8:00 AM. Stop made at Elko, Nevada, for fuel at 10:10 AM. Lunch at 12:30 and dinner at 5:30 PM. A day of rest and relaxation for the staff. Arrive Portola, California, at 5:15 PM for all night layover.

November 17, 1947
Leave Portola at 9:00 AM after arrival of Blue Ribbon Special from San Francisco carrying our guests for the trip down Feather River Canyon. Here begins the activity of the movie men who have been with us from Salt Lake on. They are assisted by an outside crew with portable equipment who follow us on the highway paralleling the tracks. Mr. Merle Johnson has crossed fingers in hope of good weather. After 11:00 AM it becomes cloudy, remaining so until we reach Oroville at 2:10 PM. We have just emerged from the canyon with its 31 tunnels and 189 miles of scenic America. Here at Oroville we are met by members of Sacramento press and radio who return with us to their city. This part of the trip is through the beautiful Sacramento Valley. Arrive Sacramento 4:45 PM. Guests detrain and we proceed to WP shops for the night. Bill Hamilton missed a very lovely trip. Seems it was very "foggo" over Chicago. Should be a moral here somewhere.

November 18, 1947
Exhibit at Western Pacific Station. Very nice crowd of Sacramentos turned out.

November 19, 1947
Leave promptly at 8:00 AM for Stockton. Chartered bus for train staff transportation to yards, runs neck and neck with T.O.T. to save us all from falling into Daniels class. Arrive at Stockton 9:30 AM. Exhibit 11:00 AM till 9:00 PM.

November 20, 1947
Leave Stockton at 3:08 PM after arrival of Oakland and San Francisco press on WP at 2:40 PM.

November 21–24, 1947
Exhibit at 47th Ave. & 14th St., Oakland.

November 25, 1947
Leave exhibition site at 11:30 PM (24th) and proceed to San Francisco via Southern Pacific rail, arriving at 3rd and Townsend SP depot at 2:00 AM. Skeleton crew aboard as most of the staff cross Oakland Bay Bridge by car in the afternoon of the 24th. Luncheon 12:15 St. Francis Hotel, Mr. H. L. Hamilton is speaker. Exhibit starts 2:00 PM at SP Depot.

November 26–30, 1947
Exhibit at SP depot, 3rd and Townsend. Crowd moves slowly but perhaps this wonderful California sunshine does slow one up. Mrs. Fay Marvin, wife of Detroit Diesel representative staff member, pays visit at this point.

December 1st and 2nd, 1947
The T.O.T. moves north today toward the country visited so long ago by Lewis and Clark. Our Mr. Clark of the train staff travels somewhat differently as we leave San Francisco at 9:00 AM. After a short run down the peninsula we cross the bay and at 1:00 are back in Oakland. We head for Davis on the main line of the SP and here we branch off to the north toward Gerber and Redding. As we approach Redding we are in sight of Mt. Shasta, covered with snow although the valley through which we are traveling is warm and bright. Both lunch and dinner (roast duck) are served in the Astra Dome. Soon after we pass Dinsmuir at 7:20 PM there is a forty-car wreck in the yards. Traffic is tied up for hours but we miss the delay by minutes. Slow schedule eats a lot of fuel and the steam generator must be shut down to get us into Klamath Falls for refueling at 11:30 PM. Reach Eugene at 7:35 AM and wait in yards until 12:25 noon. Arrive Salem at 2:15 PM for press pick up and visit by Governor Hall. Arrive in Portland at 4:15 PM. Wonderful advance publicity evidenced by radio and press. Train Director Hamilton came to San Francisco by train so was present for this leg of the trip.

December 3–6, 1947
Exhibit at Union Station. Staff installed at Multnomah Hotel. Good food at "The Broiler." Everyone has taken the trip up the Columbia River Highway to Bonneville Dam and other points.

December 7, 1947
Leave Portland before daylight at 7:00 AM and by 7:30 Inspector Schneiderbauer has the perfect Sunday morning breakfast ready. Stops made at Kelso and Centralia to take on parties of 2 and 4, respectively. Press boards at East Olympia for ride into Tacoma. Very nice crowd in charge of Ken Youel and John McGinnis. Leave East Olympia at 9:55 AM and arrive in Tacoma at 10:50 AM. Display starts at 12:30 noon. Slight rain keeps crowd small.

December 8, 1947
Public showing at Union Station Tacoma, 2:00 PM till 9:00 PM. Cold Dark Day.

December 9, 1947
Leave Tacoma at 9:45 AM with party of Seattle press just arrived on the southbound train. Short run to Seattle seems to impress Seattleites that "we have something here." Arrive 11:00 AM Northern Pacific Station. Luncheon at Olympic Hotel addressed by Paul Garrett in one of his best talks.

December 10–12, 1947
Exhibit at Occidental and Connecticut Ave. yards. Weather damp and gloomy. Crowd enthusiastic but very slow moving.

December 13, 1947
Back out of the yards to Y train at 12:30 AM for the long trip south to Los Angeles. Breakfast, lunch, dinner and a second breakfast and lunch on this hop. This is our longest single jump to date. By breakfast time Saturday morning we are past Portland and climbing up into the snow-covered mountains. Crescent Lake at the top of the long climb is beautiful and Lake O'Dell is in view as lunch is being served. Another field day for staff camera fans. Stops made at Eugene and Klamath Falls for fueling. At dusk we are crossing the California state line with Mt. Shasta straight ahead. Another stop for fuel at Gerber, 11:15 PM.

December 14, 1947
Arrive San Jose 5:35 AM, to await arrival of Mr. Mercier, SP president who is to ride with us to Santa Barbara. Leave San Jose at 9:15 AM for a beautiful trip through Santa Clara Valley and down the SP Coast line, route of the famous "Daylights." Stop made at San Luis Obispo for fueling at 1:25 PM. Large crowd on station platform. At 4:30 we arrive at Santa Barbara and are met by Don Still and quickly transported to the Santa Barbara Biltmore where a very fine reception is held. Mexican motif. Thanks, Don, it was appreciated by everyone.

December 15, 1947
Los Angeles press group entertained at Biltmore for lunch and at 1:50 PM are picked up at crossing near hotel for trip into Los Angeles. Arrival at Glendale 5:00 PM and at Union Station 5:30 PM. This show has been advanced by both Frank Harting and Ed Fish (composer of "The Beautiful Train Of Tomorrow"). These boys have done a grand job with advance publicity work, each working every other city. For them, moving day was a nightmare of boxes, trunks, and baggage.

December 16, 1947
Leave Union Station for Glendale at 12:40 PM and after much picture taking and finger pointing we leave Glendale at 2:20 PM with a large group of movie stars, GM executives, and press agents. This is to be known as the "Celebrity Ride." The chair car, "Star Dust," is justifying its title. Trip to Saugus and return by 4:30 PM was a heyday for the photographers, both pro and amateur. Catering service by Giro's and some very skillful engineering in the cab of the locomotive by Mrs. Paul Garrett and Miss Audrey Totter helped make this a very unusual trip. Mr. Rukgaber's bride-to-be, Miss Doris Shotwell, of New York City, was present and receiving best wishes from everyone. Best Wishes from all of us, George and Doris.

December 17–22, 1947
Exhibit at Exposition Blvd. at Hoover. Wonderful weather plus unusually large crowds made this a very satisfying show. Top attendance on Sunday, December 21, was 14,428.

December 23–27, 1947
Train under wraps at Southgate Plant. Train staff at their homes in the East, and the Rukgabers honeymooning in Mexico. Mike O'Brien spends his holiday on a "camera cruise" to Lake Arrowhead.

December 28, 1947
T.O.T. leaves Mission coach yards at 8:30 AM to Glendale Station. At 9:30 AM leave for Saugus with the University of Michigan football squad aboard. Return to Glendale at 2:15 PM. Train turned at Mission Coach yards and departs at 4:00 PM for San Francisco to meet the U. of M. band.

December 29, 1947
Arrive San Francisco at 4:00 AM. Leave 10:00 AM with band of 149 pieces aboard. Arrive Glendale 8:45 PM. Train returned to coach yards.

December 30–31, 1947
Train given thorough cleaning at coach yards. No more movements for the year 1947. Best wishes to a Happy New Year.

January 1, 1948
Happy New Year! T.O.T. leaves coach yards at 2:00 AM for Santa Barbara. Arrive 5:00 AM and leave at 9:00 AM with University of Southern California football team on the way to the New Years day Classic, the Rose Bowl Game. Arrive Glendale at 11:00 AM where team and train staff are transported to Pasadena. Staff members are guests of U.S.C. at the game. Train turned over to Santa Fe R.R. at Downey Jct. The Southern Pacific has been our host since November 25th.

January 2, 1948
Leave Los Angeles at 1:00 PM for trip to San Diego. Ten winning contestants of the "What I Like Best About the Train Of Tomorrow" contest aboard. Arrive San Diego 5:00 PM

January 3–4, 1948
Public Exhibit at Santa Fe Station S.D.

January 5, 1948
Leave San Diego 8:10 AM to arrive Los Angeles, Union Station at 11:00 AM. Train moved to Santa Fe yards for cleaning.

January 6, 1948
Leave Los Angeles at 8:05 AM in heavy fog. Arrive Riverside 9:45 AM. Showing 11:00–2:00 PM. Press arrive from San Bernardino and at 3:30 we are rolling the short 12 miles to that city. Showing 5:00 PM–9:00 PM. Exceptionally large crowd. Last person leaves train at 11:30 PM.

January 7, 1948
Leave San Bernardino at 12:01 AM for night ride to Phoenix. Stop at 10:00 AM Wickenberg, Arizona, for 5 hours layover. Here we are greeted by local citizens staging an old-fashioned train robbery. Masked "bandits" on horseback abduct our Manager and Engineer. Press group from Phoenix arrives 2:30 PM. Arrive 4:45 Phoenix. This is to be our last show in the territory of Don Still and Ray Hayes. It has been a lovely two months. The number of people who showed up at San Bernardino showing would indicate that Ray must have promised free drinks in his advance arrangements.

January 8, 9 & 10, 1948
Exhibit at Union Station Phoenix. Temperature 80 (degrees)

January 11, 1948
Leave Phoenix at 12:05 AM for Albuquerque. Stop made at 7:10 Williams to pick up those train staff fortunate enough to visit Grand Canyon the day before. Stop over at Flagstaff to make way for the Super Chief. Soon we are under way again and on one of the best tracks in America, with Road Foreman Osburn of Santa Fe at the throttle. The T.O.T. is given a speed test between mileposts #328 and #315. Clocked mile shows 26 seconds (138 miles per

hour), although speedometer registers 118½. Showing at Winslow 10:00–12:00 AM. Press pick up at Grants 3:45 PM. Arrive Albuquerque at 6:30 PM

January 12, 1948
Exhibits at Santa Fe Station, Albuquerque 10:00 AM–9:00 PM. Jack Benny and party visit T.O.T. while their train, the California Ltd., stops for 15 minutes on its way east.

January 13, 1948
Leave Albuquerque via Santa Fe at 9:30 AM for El Paso. Press pick up at Rincon 3:00 PM. Arrive El Paso 5:00 PM

January 14–16, 1948
Exhibit at Union Station El Paso.

January 17, 1948
Depart from El Paso via T&P lines 12:30 AM. Arrive Big Springs 8:30 AM. Exhibit 12:00 noon until 9:00 PM. No long line but excellent attendance. Weather clear and cold.

January 18, 1948
Train boarded at midnight after most successful small town showing. Leave Big Spring at 5:30 AM, arrive Sweetwater 7:15 AM. Showing 8:00 AM–10:00 AM. Depart immediately for Abilene, arriving 11:30. Showing 12:00 noon until 9:00 PM

January 19, 1948
Leave Abilene 12:30 Noon and arrive Millsap 3:00 PM for Dallas PM press pick up. From Abilene to Cisco we are hosts to several Chevrolet folks reminiscing with Mr. J. E. Johnson, who spent a number of years in this region. Arrive in Dallas 5:00 PM.

January 20–24, 1948
The "unusual" snow has blanketed Dallas. This is much in contrast to the warm reception accorded us by Mr. Frank Harting, who welcomed us to his home town. Exhibit at Dallas Fair Grounds.

January 25, 1948
Staff boards train at 10:00 PM (24th) after the show. Small reception held in observation car for the Rukgabers. Gifts and "trimmings." Depart Dallas 4:00 AM over "Katy" line for Austin. Arrive Georgetown 8:30 AM for press pick up. Group in charge of George Currier. Dormitory car in consist of train prevents many of our guests from visiting the locomotive. Arrive Austin at 9:30 AM. Rainy. Exhibit 12:00 PM–9:00 PM at "Katy" station.

January 26, 1948
Full exhibition day at "Katy" station in Austin. "Rin Tin Tin," famous movie canine, visits in forenoon and, after a few pictures, places his "stamp of approval" on the skirt of the locomotive.

January 27, 1948
Leave Austin at 3:30 PM with press group from San Antonio, just arrived by special train. Ice coating on trees, rails, and even some on the train. Arrive San Antonio, Katy station at 5:40 PM. The city "where the sunshine spends the winter."

January 28, 1948
Depart from Katy station at 6:30 AM Fort Sam Houston. Train displayed to patients of Brooke General Hospital there. Witness the results of another ice storm. Train staff hosting the train for this occasion consider themselves only slightly more fortunate than the boys confined to the hospital. Temperature at 20 degrees and strong wind. Show from 8:30–11:15 AM. Leave Fort at 11:40 and return to San Antonio. Open for luncheon guests at 2:00 PM.

January 29–31, 1948
Exhibit at Katy station in San Antonio. Weather very cold.

February 1, 1948
Sunday a free day, used to do the many jobs which are impossible to perform while the train is on exhibition.

February 2, 1948
On the stroke of 11:00 AM we leave San Antonio behind and are making our way up the "Katy" tracks for Fort Worth. The press pick up is to be made at Waco and are due there at 3:30. Arrive at Waco 3:15 PM allowing our manager, Mr. Rukgaber, to visit the Freedom Train showing in Waco today. We are due to leave at 3:45 but delay our departure until 4:05 to allow us to escort Mr. O'Brien, manager of the Freedom Train, through the T.O.T. Arrive Fort Worth Union Station at 6:00 PM

February 3–7, 1948
Exhibit at Union Station. Cold and wet.

February 8, 1948
Leave Fort Worth at 4:30 AM on Texas and New Orleans lines. Staff boards train the night before at 10:00 PM. Press pick up at College Station, 9:45 AM for ride into Houston, arriving 11:20 AM.

February 9–13, 1948
Showing at Southern Pacific Station.

February 14, 1948
Leave Houston at 5:15 AM on T&NO lines. Arrive Logansport at 10:20 for press pick up. Arrive Shreveport 11:30 AM. Showing 1:00 PM–9:00 PM.

February 15, 1948
Leave Shreveport 3:10 AM on Kansas City Southern lines for Baton Rouge. Press pick up at New Roads 9:40 AM arriving Baton Rouge at 11:30 AM. Showing 1:00 PM–9:00 PM at Illinois Central Station on the banks of the Mississippi. Hotel Heidelberg only one-half block from site. Johnny Duss plays Santa Claus to taxi driver by requesting a ride to IC station which proves to be on the other side of the building.

February 16, 1948
New Orleans press group arrive at 3:44 PM for return trip into city on T.O.T. On the Illinois Central road. The train staff has spent the morning seeing (and hearing) the works of Hughie P. Long. Leave Baton Rouge 3:50 and arrive New Orleans at 5:30 PM.

February 17–20, 1948
Exhibit at New Orleans, IC Station Herschel Willis, court jester of the train crew, meets with the "N.O. Trainman" to compare notes.

February 21, 1948
Leave New Orleans at 7:00 AM for trip to Jackson, Mississippi. Press pick up at Goodhaven 9:30 AM. Arrive Jackson at 10:30 AM in cold wet rain. Train site about one mile from station. Showing 1:00 PM–9:00 PM.

February 22, 1948
Leave Jackson 12:01 AM on IC lines as far as Meridian. Switch to Gulf, Mobile and Ohio road. Proceed to Mobile, arriving 7:20 AM. Sunday showing on a cold damp day 10:00 AM–5:00 PM. Leave Mobile on Louisville and Nashville road at 7:00 PM for Jacksonville.

February 23, 1948
Arrive Jacksonville at 8:45 AM. Switch to Florida East Coast road and depart for Miami at 10:50 AM. Stop made at Vero Beach 2:24 PM for Mr. and Mrs. C. E. Wilson and party. Arrive Miami at 5:30 PM.

February 24, 1948
To take the place of a Miami press ride two round trips are made to Palm Beach on this date. First group, mostly press and radio people, leave Miami at 9:15 AM and return 12:15 PM. Second group, including Mr. Wilson and Mr. Kettering, leave Miami at 2:10 PM, returning 5:30 PM.

February 25–26, 1948
Showing at 15th Road and Brickell Ave.

February 27, 1948
Evening ride in charge of Mr. Paul Garrett for business and social leaders vacationing in the vicinity. Left Miami 11:30 PM for trip to Homestead and return. Arrive back in Miami at 2:00 AM.

February 28–29, 1948
Showing for these days include morning shows, manned by local Boy Scouts as an experiment. Highly satisfactory for school group showings.

March 1, 1948
Leave Miami 11:30 PM (29th) on FEC lines to Benson Jct. Here we are switched to Atlantic Coast Line for remainder of trip to Tampa. Press ride pick up at Haines City 9:30 AM arriving Tampa 11:15 AM. Showing 1:00 PM–2:00 PM for ticket holders. 2–9 Public.

March 2–4, 1948
Our first show "under new management." Show at Cass and Jefferson, Tampa. Goodbye, George Rukgaber, you were tops, and "Good Luck" to you, Roland R. Seward, we're for you.

March 5, 1948
Leave Tampa at 8:30 AM for Orlando, pick up press at Haines City 9:30 AM arriving Orlando 10:45 AM. Showing at 2:00 PM. Mr. Ernest Dunsford, father of Train Business Manager, aboard this trip.

March 6–7, 1948
Showing at Hughie Ave. & Gore Street. Light rain most of the days. Some movie shots made here Sunday morning the 7th by Sound Masters, Inc.

March 8, 1948
Leave the depot at Orlando 12:01 AM for Montgomery. Goodbye to Florida and the sun. Atlantic Coast Line road to Albany, Ga., and Central of Georgia from there to Montgomery. Arrive Columbus, Ga., 10:45 AM EST. Leave Columbus at 1:25 PM CST for Montgomery. Press pick up at Union Springs 3:15 PM. Arrive Montgomery 4:30 PM.

March 9–10, 1948
Exhibit Union Station. Rain. John J. Batsche of Frigidaire returns to Dayton for a breather. He is substituted for by "How in the World are you," MacMillan.

March 11, 1948
Staff boards train on night of 10th for early departure (5:00 AM) for Memphis. Breakfast served aboard without the watchful eye of Joe Schneiderbauer, who for the first time in many years of railroading failed to catch his train. Much speculation as to the reason. Tough luck, Joe. Layover in Birmingham from 9:00 AM to 11:00 AM where we are switched to Frisco Lines for remainder of trip. Press pick up at New Albany 3:10 PM. Arrive Memphis 4:30 PM

March 12–14, 1948
Exhibit on Illinois Central tracks near old station on Front street.

March 15, 1948
Depart from Central Station at 8:30 AM with group of prominent guests including Judge Kelly (feminine) of Memphis and the GM Audit Department's glamour boy, Richard Radke. Press pick up at Brinkley 10:20 AM arriving Little Rock 11:30 AM. This run made on Rock Island Lines. Rain.

March 16–17, 1948
Exhibit at Rock Island Station, Little Rock.

March 18, 1948
Leave Rock Island Station at 8:38 AM. Departure time to have been 8:30, but train held for staff member who became too ill to accompany us. Mr. Wm. J. Davidson has among his guests Mr. and Mrs. Lou Valentine (Dr. I.Q.). Breakfast and lunch served. Guests detrain at Ola 10:20, press pick up at Holdenville 3:20 PM. Arriving Oklahoma City 5:05 PM.

March 19–22, 1948
Exhibit at Union Station. Nice clean side and fine crowd.

March 23, 1948
Depart from Oklahoma City at 2:05 PM on Frisco Lines for Tulsa. Press pick up at Chandler 3:10 arriving Tulsa at 4:55 PM. Large group of press and radio aboard.

March 24–26, 1948
Exhibit Union Station, foot of Boston Ave. Mr. J. E. Johnson made a Chief of the Otoe Indians Tribe at appropriate ceremony on the afternoon of the 24th.

March 27, 1948
Leave Tulsa at 10:00 AM on Santa Fe road with small group of Tulsa citizens riding as far as Bartlesville. Tulsa group detrains at 11:00 AM and at 11:30 we depart with another small group riding to Independence. Arrive 12:30 PM and depart at 1:30 PM. No guests. Lunch consists of Schwartzbergers and coffee. Stop made at Winfield 2:40 PM for Wichita press. Arrive Wichita 4:30 PM. Fine site and finer weather. Showing 6:00 PM–9:00 PM. Tickets.

March 28–30, 1948
Public showing at Old Rock Island Station on East Douglas Ave.

March 31, 1948
Free day! Mike O'Brien, John Batsche, Johnny Duss, and Pete Fleming change "excites" in coach engine, while rest of the staff imitate Admiral Byrd by having an outdoor picnic. As yet, no pneumonia reported.

April 1, 1948
Leave Wichita 8:30 AM for short trip to Newton where we have a 3-hour showing. This is the home of service manager C. W. Perkins. A total of six thousand people visited the train including many school children. Considering there was no real publicity it would seem that Mr. Perkins, his family, and friends certainly did noise it around. Leave Newton at 1:20 PM. Press pick up at Emporia 2:40 PM. Governor Carlson among those present. Topeka at 4:30 PM

April 2–3, 1948
Exhibit at Santa Fe Station, Topeka.

April 4, 1948
Leave Topeka 10:50 AM. Arrive Lawrence 11:25 AM. Exhibit at Santa Fe Station 1:00 PM–5:00 PM. Over 7,000 guests. Mr. Ed Groshell of Chicago Public Relations is on hand to familiarize himself with exhibit details.

April 5, 1948
Leave Topeka at 3:00 PM with members of Kansas City press aboard for trip into Kansas City. Arrive Union Station 4:30 PM and train is spotted at 20th and McGee for showing beginning at 7:00 PM.

April 6–10, 1948
Exhibit at 20th & McGee. Night of 10th we move to Kansas side of the river for one-day showing.

April 11, 1948
Exhibit Fairfax Food Terminal 10:00 AM–9:00 PM. Move back to Union Station.

April 12, 1948
Free Day. Showing for special party only.

April 13, 1948
Depart for K.C. Union Station at 11:45 AM for St. Louis on Missouri Pacific Lines. Press stop at Hermann 3:10 PM. Arrive St. Louis 5:00 PM.

April 14–18, 1948
Exhibit Union Station track #2. Harry Blair's party on the 14th was highlight of our St. Louis visit.

April 19, 1948
Early departure from St. Louis Union Station at 7:00 AM over Burlington Route to Burlington. Tracks follow Mississippi River most of the way. Arrive 12:30 PM. Showing from 5:00 PM–9:00 PM. This exhibition advanced entirely by the fabulous Ellsworth E. Seitz was a fine success.

April 20, 1948
Leave Burlington 8:30 AM on Rock Island Lines. Stop at Cone for the parents of Gaylor M. Coffin, Public Relations Executive, and stop made at Waterloo 12:10 PM for ex-train Manager Rukgaber. Press pick up Albert Lea 3:00 in charge of J. F. Fitzgerald. Minneapolis 5:30 PM

April 21–25, 1948
Showing at Fair Grounds, midway between St. Paul and Minneapolis. Fine crowds and some rain. This show on Great Northern rails.

April 26, 1948
Depart from exhibition site at 8:00 AM. Switched to main line and leave St. Paul yards on Chicago & Northwestern lines. Beautiful ride through lovely country brings us to Beaver Dam at 3:30 PM to wait signal that party of press has arrived at Friesburg by Greyhound. Drivers apparently confused by absence of street cars and traffic cops, so it's 4:10 when they arrive at Friesburg with party. Short run from Beaver Dam to Friesburg for pick up. Arrive 6:10 PM at Milwaukee.

April 27–30, 1948
Exhibit at Chicago & Northwestern Station.

May 1, 1948
Leave Milwaukee at 10:10 AM with members of Janesville and Madison press arriving on C&NW train. Stop at Madison to let Madison press detrain. Proceed to Janesville arriving at 12:50 PM. Showing 2:00–9:00 at Chevrolet plant.

May 2, 1948
Show at Chevrolet plant. Rain.

May 3, 1948
Leave Janesville at 8:01 AM. (That minute charged to Howard Dunsford.) Arrive Madison 9:00 AM. Mr. and Mrs. Ed Groshell of Public Relations aboard.

May 4, 1948
Exhibit at C&NW Station, Madison.

May 5, 1948
Free day at Madison. Depart 11:00 PM for Des Moines via Chicago.

May 6, 1948
Arrive Chicago yards at 3:00 AM and move into LaSalle Station at 8:30 for 9:25 AM departure for Des Moines. Small group GM officials aboard. Pick up party at Bureau, Ill., including guests of Mr. J. D. Farrington, Rock Island president. Lunch served to 150. Guests detrain at Davenport and Morengo, Ia., where short ceremony takes place. Gov. Blue and Mr. J. E. Johnson speakers. Press pick up here. Arrive Des Moines 6:00 PM.

May 7–9, 1948
Exhibit at Rock Island yards, Des Moines.

May 10, 1948
Leave Des Moines 7:30 AM for Davenport. Press pick up at Iowa City 9:30 AM arriving Davenport 11:00 AM. Threatening rail strike curtails exhibition here to luncheon guests only and at 3:00 PM we are on our way into Chicago, where T.O.T. is to undergo an overhaul. Arrive Blue Island 7:00 PM. After much switching and delay we are safe in Electro-Motive Plant #2 yards at 2:00 AM. Fine ride to 63rd St. for part of train staff in a Rock Island caboose.

Plant City Tour
May 11–28, 1948
    Shopping period. New paint inside and out as well as many mechanical corrections.

May 29, 1948
Special showing for Electro-Motive employed and families on open house day at 103rd Street plant.

May 30, 1948
Leave plant yards at 9:30 AM for move to Dearborn Station. Start at 2:30 PM for Danville, Ill. This is first move of plant city tour. Mr. J. M. Budd, president of Chicago and Eastern Illinois, on whose road we are making this trip, is aboard. Press pick up at Chicago Heights 3:20 PM. Arrive Danville 5:00 PM.

May 31–June 1, 1948
Exhibit at C&EI freight yards, Danville. Picture taking run of 50 miles made on June 1, 8:00 AM–11:00 AM.

June 2, 1948
Leave Danville 7:00 AM for trip to Indianapolis via NYC. Wait in Crawfordsville 8:00–8:30 AM for buses bringing Indianapolis press. Arrive Indianapolis 10:00 AM and move to Allison plant on B&O tracks. Curved track leading into shipping building leaves locomotive and two cars inside and two cars remaining outside. Exhibit 1:00 PM–8:00 PM.

June 3, 1948
Allison Plant.

June 4–5, 1948
Leave Allison Plant at 9:30 PM (night of 3rd) for Chevrolet Plant. Exhibit 10:00 AM–7:00 PM.

June 6, 1948
Free day. Work completed on new generator set for dormitory car.

June 7, 1948
Leave Chevrolet 11:00 AM for switch to Union Station. Depart with Anderson group aboard at 2:30 PM. Arrive via NYC at 3:15 PM amid terrific hail storm. Astra domes withstood impact of hailstones baseball size. Showing at 5th & John St., Anderson. Open 7:00 PM–9:00 PM this day.

June 8–13, 1948
Exhibit at 5th & Johns Street. Visits to Guide Lamp and Delco Plants and a luncheon at the Country Club for the train staff makes our week most enjoyable.

June 14, 1948
Leave Anderson 10:30 AM via New York Central to Muncie. Press Budd train before departure from Anderson. Arrive Muncie 11:45 AM. Show 2:00–9:00.

June 15, 1948
Exhibit at Nickel Plate yards, Muncie.

June 16, 1948
Leave Muncie via Nickel Plate road for Ellwood at 10:00 AM. Press boards here and we switch to Pennsylvania Lines for remainder of trip to Kokomo. Arrive 11:30 AM. Show 2:00–9:00 PM. Buffet luncheon held at Francis Hotel. No speeches.

June 17, 1948
Exhibit Pennsylvania Station, Kokomo.

June 18, 1948
Leave Kokomo 5:00 AM for Bedford via Penna. Switch to Monon Route at Greencastle 10:30 AM and pick up press group. Arrive 11:45 AM. Indiana Limestone Centennial gives festive air to Bedford. Show 2:00–9:00 PM.

June 19, 1948
Leave Bedford 15 minutes ahead of scheduled time of 12:30 AM for Michigan City via Monon. Clifford Mezey, new train staff member, is "left at the post" because of time change. Emergency arrangements put him on the train at Bloomington, thirty miles away at 1:10 AM. Arrive Michigan City 6:30 AM. Showing 9:00 AM–1:00 PM with Pullman Standard men in charge of Preston Calvert acting as hosts. Bill Lewis site manager for this stop. Fine job, Bill. Leave 1:45 PM via C&O route for Grand Rapids. Press pick up at Bangor 3:15 PM. Arrive G. R. 5:15 PM. Show 8:00–9:00. A new high in Regional Manager courtesies was the presence of Hotel Rowe representative to handle train staff luggage. Thanks, Walter Scott, for that and the many others you have shown.

June 20–24, 1948
Showing 5:00–9:00 PM Sunday (20th), other days 2:00–9:00 PM at Grand Trunk Station. Staff picnic on Sunday.

June 25, 1948
Leave Grand Trunk Station 9:00 AM for Lansing via Grand Trunk lines. Press pick up of 100 at 9:45 at Ionia. Arrival time in Lansing 11:30 AM. Show 2:00–9:00 PM.

June 26–29, 1948
Exhibit at Grand Trunk Station, Lansing.

June 30, 1948
Buses arrive with Flint press group at 3:00 PM and at 3:30 we leave for Flint on the Grand Trunk Line. Arrive Flint site near IMA Bldg. at 5:00 PM. Press buffet supper at Flint City

Club for guests. Among those present were Mr. C. S. Mott and Mr. Harlowe Curtice. Showing 7:00 PM–9:00 PM for IGA girls.

July 1–5, 1948
Showing 2:00–9:00 at IMA site. Each day a service club of Flint were given a short talk about the train by Train Manager Seward at their weekly luncheon meeting and this is followed by a special showing of the Train. One such club was shown a movie of the train's western tour, narrated by Max Wilson of the train staff. Jack Batsche and Johnny Duss try the Michigan fishing which is so popular in these parts.

July 6, 1948
At 9:30 AM we leave Flint with small group of Detroit office people for Lake Orion. This move takes us over the Grand Trunk road to Lapeer and then NYC to Orion. Flint to Lapeer is run in reverse because of the absence of a Y at Lapeer. Arrive at Orion 11:00 AM. Here we are guests of Mr. Lee Anderson, GM dealer, and show site is near his new ultra-modern showroom. Leave Orion 6:00 PM in reverse to Lapeer and then to Flint, forward, arriving 7:30 PM.

July 7–11, 1948
Show IMA site, Flint. No long lines but a steady appreciative crowd.

July 12, 1948
Leave Flint 10:00 AM for Saginaw on C&O R.R. Saginaw press group aboard. Short ride to Saginaw is made longer by much switching to arrive at site on west side of Saginaw River near Michigan Bean Elevator. Saginaw plants are hosts at luncheon with Mr. Nelson C. Dezendorf as speaker. Showing 2:00–9:00 PM. Fine view to T.O.T. from Genessee Ave. bridge.

July 13–16, 1948
Showing at riverside site 11:00 AM–9:00 PM. Best turnout in months.

July 17, 1948
Leave Saginaw site at 8:30 AM with members of Bay City press aboard. C&O route takes us to Flint and back to Bay City. Arrive 11:45 AM at our site in the C&O Station. Show 2:00–9:00 PM. Roland R. Seward, Train Manager, speaker at plant luncheon.

July 18, 1948
Showing noon till 9:00 PM. Walter Scott scores another "first" as host to the train staff on a fishing party Sunday forenoon.

July 19, 1948
Leave Bay City site at 11:50 PM for the Chicago Railroad Fair on the Chicago Lake Front, via Grand Trunk Lines. Arrive Harvey, Indiana, 5:15 AM and are taken to Fair site by Illinois Central, arriving 6:30 AM on Track #2 where we hope to be the #1 exhibit. Fair dates July 20–September 24, 1948.

July 20–Sept. 24, 1948
This period spent at Chicago Railroad Fair on Chicago lakefront. Some of the highlights of this exhibit were a television broadcast, visits by many famous people, and a daily attendance which exceeded our most fond hopes. Average daily attendance in excess of 14,000 people at T.O.T. One of the problems to overcome was the absence of good eating accommodations for both Electro-Motive personnel, acting as hosts, and our own staff. This was bridged by serving sandwiches and coffee twice a day from our own diner. Everyone voted this an ideal solution. This was also a period of vacations for the staff on an alternating schedule. Burns Agency guards protected the Train at all times.

The only movement during this period was on September 8th when we left the grounds for one day only to carry the Dayton press group to Indian Lake and return to Dayton. Left Chicago 12:01 AM arriving Dayton 10:00 AM.

September 8, 1948
Leave with guests at 3:00 PM arriving Indian Lake 5:30 PM. Return trip started at 9:30 PM arriving back at Dayton 11:45 PM and leave at once for Chicago. We are switched into Fair Grounds at 9:00 AM the 9th. Entire trip made via New York Central except the return to Dayton from Wapakoneta on the Baltimore and Ohio. 224 persons on the ride made moving about on the train somewhat of a problem, especially in the lounge car.

September 25, 1948
Leave Fair Grounds at 2:01 AM after much switching to clear exhibits on track behind us. Proceed to Harvey and by 5:00 AM are on the Grand Trunk headed for Detroit. Arrive Brush St. Station 11:40 AM where Train is cleaned for opening at 9:00 the following day.

September 26, 1948
Showing at Detroit Transmission Plant, Riopelle and Farnsworth. Special showing 9:00–12:00 for American Bankers Association and wives who are GM guests for the day.

September 27–29, 1948
Central Office showing at Second and Sears.

September 30–October 1, 1948
Chevrolet Gear and Axle showing at Holbrook and St. Aubin.

October 2–3, 1948
Cadillac open house showing, Michigan and Clark. Cadillac neckties and lighters to train staff.

October 4, 1948
Courtesy ride on Grand Trunk to Pontiac and return for GM executives. Leave Brush St. Station 3:30 PM, returning 5:30 PM. Cadillac Motor is host to train staff at D.A.C. luncheon.

October 5–7, 1948
Exhibit at Ternstedt Plant, Fort and West End Ave., Detroit.

October 8, 1948
Show at Fisher Body Plant, Piquette & Hastings.

October 9, 1948
Exhibit Detroit Diesel Plant, Telegraph and Plymouth. Fay Marvin is host to train staff in the evening.

October 10, 1948
Leave Diesel Plant 7:00 AM via C&O and by switching to Grand Trunk at Milwaukee Jct. arrive Pontiac 10:05 AM. Site, opposite Pontiac Plant.

October 11–14, 1948
Exhibit at Sanderson and Oakland Ave., Pontiac.

October 15, 1948
Leave Pontiac GT Station 10:00 AM, arriving Union Depot Detroit 12:45 PM. Toledo press arrives for departure at 2:30 PM and arrive Toledo 4:45 PM via C&O. Gay crowd.

October 16–17, 1948
Exhibit at foot of Washington St. on B&O tracks. Steam generator working for first time of the season. Bill Hammond (The Senator) of Detroit Diesel is a very welcome addition to our staff. He certainly knows his equipment.

October 18, 1948
Leave Toledo via NYC for trip to Cleveland at 12:55 PM, arriving at 3:05 PM to wait arrival of press group from Warren. Leave 55th St. Station for Erie at 4:00 PM and proceed to

Warren on the Erie line, arriving 5:10 PM at site. Train staff quartered at Pick-Ohio, Youngstown.

October 19–21, 1948
Exhibit Erie freight yards, Warren. Mornings and early afternoons given over to schools who arrive and depart via school bus. Excellent showing.

October 22, 1948
Leave Warren 10:00 AM via Erie for Buffalo. Arrive 4:30 PM and are transferred to NYC for switch to Chevrolet Tonawanda Plant, on site at 5:30 PM.

October 23–24, 1948
Exhibit at Tonawanda Chevrolet Plant. Nice site.

October 25, 1948
9:30 AM departure from Chevrolet via NYC with Lockport press aboard. Arrive 11:30 AM. Country Club luncheon as guests of Harrison Radiator Division. Mr. R. R. Seward Jr., Train manager, is guest speaker.

October 26–27, 1948
Exhibit in Lockport freight yards of NYC. Train staff at Hotel Statler in Buffalo, 22 miles away.

October 28, 1948
Leave Lockport 10:30 PM (27th) on NYC in dense fog for Framingham. Eddie Braken and bride aboard to Rochester. Arrive Springfield, Mass., at 6:20 AM. Press arrives by special train at 10:00 AM for trip to Framingham. Arrive 11:20 AM. Show 2:00 PM. Don MacShane is all set with site.

October 29–31, 1948
Exhibit B&A (NYC) freight yards. Train staff at Sheraton Hotel in Worcester 20 miles away. Country beautiful at this time of year. Cars furnished for transportation. Hallowe'en party on Saturday nite.

November 1, 1948
Leave Framingham at 1:10 PM for Springfield, arriving 2:30 PM. Press from Meridan, Conn., board at 3:20 and we are soon on the NYNH&H bound for Meridan. Arrive 4:30 PM. Jim Gleason has cramps from greeting old friends. Bond Hotel in Hartford for staff.

November 2–4, 1948
Fine site for exhibit in freight yards near International Silver plant. I.S. made a generous offer to clean all our silverware, which was promptly accepted. New low in hotel accommodations, at Red Cedars cabins, four miles out. We're slipping.

November 5, 1948
Leave Meridan 8:00 AM after train is turned at New Haven. Large party of Plant people aboard for ride to Bristol via Hartford on NYNH&H. Show 11–9.

November 6–7, 1948
Exhibit at Forrestville, two miles from Bristol. Fine crowds. Staff at Endee Inn, a division owned hostel. Let's stick to Ball Bearings.

November 8, 1948
Leave Bristol at 10:05 AM on the NYNH&H road for Elmsford, N.Y. Change to NYC at Brewster 12:30 PM. Layover here with 4:00 PM departure time is made even later [by] arrival of buses with press party. Leave Brewster at 4:40 PM, arriving Elmsford 5:40 PM in darkness. Staff in White Plains, 3 miles away. Tarrytown plant two miles in opposite direction.

November 9–10, 1948
Exhibit at NYC Elmsford Station. Very rainy weather. Steady crowd.

November 11, 1948
Depart from Elmsford at 12:01 AM for trip up the Hudson and across the river at Albany. Familiar territory for the T.O.T. Arrive Weehawken 9:30 AM and are switched to Erie at this point for balance of trip to Bloomfield, N.J. Arrive 11:30 AM. No show today. Staff at Sheraton Hotel in Newark, N.J. Pete Fleming is a little short this morning. Choir practice last nite.

November 12, 1948
Exhibit at Erie Freight Yards, Bloomfield, N.J.

November 13, 1948
Harrison, N.J.

November 14, 1948
Leave Harrison 1:10 AM via Penna. for Linden Plant of BOP. Arrive 3:00 AM. Special visit by group of Sisters from South Amboy and Red Bank, through the courtesy of Walter Scott of Flint. Lovely weather and very long line during the afternoon.

November 15, 1948
Leave Linden Plant at 3:00 AM arriving at the New Brunswick plant of Delco Battery Division at 4:00 AM via Penna. Staff moved to Roger Smith at New Brunswick for day only. Open house at plant.

November 16, 1948
Depart from site at 4:00 AM via Penna. to Metuchen and thence to Clark (Hyatt Plant) via Lehigh Valley Road, arriving 7:30 AM. Staff at Elizabeth.

November 17, 1948
Press boards at Plant and we back out of "S" curve to main line of Lehigh at 10:30 AM. Switch to Reading road at Monville and proceed to Trenton, arriving 12:00 noon. Site on Willow Avenue yards of Reading, close to center of city.

November 18, 1948
Public exhibit at Reading freight yards, Willow Ave., Trenton. Lots of pickaninnys.

November 19, 1948
Staff boards train at 5:00 AM for Philadelphia via Reading road and then to Wilmington via B&O. Our old friend Bill Watson of B&O is with us again by our request. Arrive at plant 7:45 AM. Showing 3:00 PM.

November 20, 1948
Exhibit at BOP Plant, Wilmington, for the second time. Train is partially hosted by young ladies from BOP. Unusual but nice arrangement.

November 21, 1948
Leave BOP Plant at 8:00 AM with group of 80 supervision people for trip to Baltimore. Arrive 9:30 AM at Camden Station. This date is the start of a fine holiday for most of train staff.

November 22–25, 1948
Train remains on siding at Camden Station. No show for this Thanksgiving week, but time is used by our service men to take care of minor repairs and cleaning.

November 26, 1948
An early morning departure at 3:45 for Columbus on the B&O road. Stop for fueling at Cumberland 6:00 AM and short stop at Pittsburgh to pick up some of train staff meeting us there at 10:20 AM. Slow schedule to Zanesville where we pick up Columbus press at 3:30 PM. Arrive Columbus 5:00 PM. Train turned and backed into Ternstedt Plant.

November 27–December 1, 1948
Exhibit at Ternstedt Plant, Columbus. Soft ground makes fueling a problem.

December 2, 1948
Leave site at 10:30 PM (12/1) after close of show, with party of ladies, representing Dayton press. One-hour delay at Union Station before we depart at 12:45 AM via NYC. Arrive Dayton 2:00 AM. Party in charge of the Chope and Culp team of Cleveland.

December 2–7, 1948
Exhibit at B&O team tracks, Ludlow St., Dayton for the second time. Employe showing every day. Public admitted because of slow crowd.

December 8, 1948
Breakfast aboard as we leave Union Station at 8:00 AM for the short trip over the B&O to Hamilton, arriving 9:30 AM. A cold morning and the water pipes previously laid for our use are frozen. Trip through the Fisher Plant for staff as well as a plant luncheon. Show at 2:00 PM on Route #4 on outskirts of city.

December 9, 1948
After the day's exhibit and the Blue Goose is loaded we leave at 10:10 PM for Dayton via B&O and switch to NYC at 11:20 PM for overnight ride to Elyria. Bill Watson leaves us here to return to Chicago until our next show on B&O lines some months away. Proceed to Elyria arriving at 7:30 AM.

December 10–11, 1948
Here we are greeted with our first snow of the season. Exhibit at Brown-Lipe-Chapin Plant west of the city. Open house here on the 11th. Many GM products on display in this new modern plant.

December 12, 1948
Early Sunday morning departure at 6:30 from B.L.C. Plant for Sandusky via NYC. Cold and rainy. This stop is our third show at a New Departure city. Welcoming ceremony ruled out because of bad weather. Show open at 11:00 AM.

December 13, 1948
Last day of plant city tour. After exhibit closes and Blue Goose is loaded we leave Sandusky at 10:30 PM via NYC for St. Louis, Mo., and the Pullman shops. Don MacShane left us today for a few days in the Detroit office, to be followed by a Christmas holiday and on January 1, takes up his new duties as Advance Representative for T.O.T. Good luck, Don!

December 14, 1948
Arrive in St. Louis Union Station at 10:00 AM for transfer to Terminal Railroad and the Pullman shops on Bircher Blvd. Here the T.O.T. will be given a face lifting in preparation for our 1949 tour to start January 17th.

December 15, 1948–January 14, 1949
This is it! The first real overhauling and refurbishing job with enough time allowed for all the necessary work. Cars were entirely stripped to allow for new upholstering, floor covering, and painting. Among the many things done during this period was the complete redecoration of the private dining room, fittings and railings of locomotive were chrome plated, painting throughout, wheels turned on all cars, and all undercar engines overhauled. The Blue Goose gets a name plate as well as a legal right to use the name, a fine job of shadow painting to make it conform in looks with the other cars. This period, too, is a chance for train staff to enjoy a part of the holidays at various times. Everyone is in accord that the work performed here was of the highest quality and much credit should go to the Pullman Co. for their fine job as well as those of the train staff in charge of the work for the many innovations worked out. St. Louis weather, too, comes in for a vote of thanks.

Southeastern Tour

January 15, 1949
Here we go again! Leave Pullman shops at 3:00 PM for Union Station. At 5:30 PM we are on the Wabash, headed for Detroit. Stops made for fuel and water at Decatur and Ft. Wayne. Arrive in Michigan Central yards at 5:30 AM (16th).

January 16, 1949
Train cleaned and shined for the trip to New York the next morning. Train staff meets with Public Relations men in the evening for briefing on the events of the New York ride. Mr. Hamilton and Mr. Groehn have missed no detail.

January 17, 1949
At 6:30 AM all men concerned are ready to receive our guests, the editors from over the country who have been invited to ride with us into New York and the GM Auto Show at the Waldorf. On this trip we have two new staff members, James Gavagan, formerly of GM Previews, and Pete Galinato, a new member of Mr. Schneiderbauer's crack crew who have served us so well. Through the tunnel to Windsor, leaving at 7:30 AM on the Michigan Central line. Soon we are on the open road across the flat Ontario countryside bound for Buffalo with the speedometer hovering near 100. Stop made at St. Thomas. At Buffalo we are switched to the Delaware, Lackawanna and Western road for the remainder of the trip. Here we pick up Eddie Braken and group of seven, including an accordionist enlisted for his help in entertaining our guests. At East Buffalo, the DL&W president, Mr. White, and party of three board. We are out to make a record run and at times the needle rests on 102 as we cross New York State and into the hills of Pennsylvania. Stops made at Elmira and at Scranton. Three meals served enroute for entire party of 110. It's dark at 7:30 PM when we draw up at Hoboken. A flat 12 hours since leaving Detroit, taking one hour and fifty-four minutes off the best previous time. New York was always quite a long hop from Detroit until today. After our party leaves us for the bus ride into N.Y. we are switched via Erie and NYC to Jersey City to await departure at 1:00 AM for Rocky Mount, N.C. Statue of Liberty is beautiful in the moonlight as we are sitting in the domes during the switching operations.

January 18, 1949
Leave Jersey City at 1:00 AM via B&O to arrive in Potomac Yards at Washington 6:00 AM. Leave at once for Richmond via Richmond, Fredericksburg and Potomac, arriving 9:30 AM to be taken over by Atlantic Coast Line for balance of trip to Rocky Mount, N.C. At 12:45 PM we are standing adjacent to the Blue Goose, which came straight from St. Louis via B&O. On this leg of our journey we have Mr. Ronald R. Seward Sr., who has joined his son, Train Manager Seward, for a brief time. Train is made ready for our first 1949 show the next day.

January 19, 1949
Draw away from Rocky Mount at 8:10 AM for Wilmington, N.C., the headquarters of the ACL, our sponsoring railroad. Press pick up at Faison, N.C., 10:00 AM. Group in charge of Bedford Culp, Cleveland Region Public Relations. Arrive Wilmington 11:30 AM. Show 6:00 PM at ACL Station. Don MacShane has left a clear trail on his first advance job.

January 20–21, 1949
Exhibit at ACL Station, Wilmington. Lovely weather and oyster feed by local GM people make this a fine start on our 1949 tour. Big hotel fire on night of 20th lends excitement.

January 22, 1949
At 12:01 AM everyone is asleep as we leave Wilmington via ACL for Columbia, S.C. Press pick up at Sumter 10:00 AM. Arrive Columbia 11:30 AM. Exhibit at siding near Fair Grounds start at 6:00 PM.

January 23–24, 1949
Exhibit near Fair Grounds east of city. Weather ideal. GMAC people have much credit due them for their handling of the local details in connection with our show. Bill Lewis is on his feet again after a brief illness. Bill is doing a fine job with site arrangements, preceding us by a couple of days at each location.

January 25, 1949
Here we have a sneaker. Leave Columbia via ACL at 4:30 AM. Arrive in Sumter 6:00 AM. Local people have prevailed on us for a three-hour show here. Small ramp used at rear of train, guests leaving via coach entrance. Open 9:00 to 12:00 noon, 5,304 persons, mostly school children. As we ran out of time with still a long line waiting, we closed domes so everyone could get inside and through the train before we had to leave as per schedule at 1:00 PM. Pick up press at Denmark, S.C., 3:00 PM and arrive Augusta 5:00 PM, ending our tour with ACL.

January 26–28, 1949
Showing at Augusta in Union Terminal under sponsorship of Georgia Central Railroad.

January 29, 1949
Central of Georgia R.R. takes us out of Augusta at 7:00 AM for Macon. Press pick up 10:00 AM at Tennille, Ga., arriving Macon 11:30 AM. Exhibit open 6:00 PM at Union Station. Our first cold weather.

January 30–31, 1949
At Union Station Macon, one of Georgia's finest cities. Good hotel and exquisite food at the Lassiter Mansion.

February 1, 1949
It's a very cold morning as we board at 5:00 AM, but we are headed for St. Petersburg, Fla. Central of Ga. takes us as far as Albany, Ga., at 9:00 AM. Here we are again with ACL and after a stop at Trilby for the press at 2:00 PM we arrive in that haven of rest for the aged, St. Petersburg, 4:30 PM. Open range throughout the state makes railroading rather hazardous. Score, one steer at 65 miles an hour. No damage to locomotive.

February 2–7, 1949
Exhibit on ACL mole, extending the length of the train into the bay. A fine site. Good crowds every day but very slow moving. Hotel Soreno overlooks the bay and St. Pete has some very nice restaurants. Fishing, golf, and some swimming make this a very pleasant stop to say nothing of the sunshine. Jim Gavagan spends his time here at the Veterans' Hospital. Good luck, Jim, hope you are with us again soon.

February 8, 1949
Goodbye to the warm sunshine of St. Petersburg as we leave via Seaboard Air Line tracks at 8:00 AM for the capital city, Tallahassee. Our route takes us through Tampa (March 1948) and very near to Jacksonville (July 1947). Press pick up at Madison 3:15, arriving in Tallahassee 5:00 PM.

February 9–12, 1949
Exhibit near airport, three miles from city. Nice site but very far out and no public transportation. Nearby schools help attendance at what would have been a very quiet show.

February 13, 1949
Free day! Sunday and no show. Picnic for staff near train site, with steak roast and ball game. Guest, W. E. Hamilton, reports a slight stiffening at every joint.

February 14, 1949
Leave Tallahassee at 6:40 AM for Pensacola via SAL to Chattahoochee, thence via Louisville and Nashville to Pensacola. Press pick up at Crenshaw 10:00 AM. Arrival time

11:30 AM. We have as our guest on this trip from Tallahassee Mr. Thomsen, president of Danish Railways.

February 15–16, 1949
Exhibit at L&N Station Pensacola. Staff visits Naval Training Station and aircraft carriers.

February 17, 1949
Another early morning move as we leave Pensacola at 6:20 AM for the long ride to Nashville via L&N. Mr. James Gleason in charge of train during short visit of Mr. Seward with his family in Baltimore. Press pick up at Lewisburg, Tenn., 3:30 PM arriving Nashville 4:30 PM. Cold weather again. No steam for Blue Goose so crew goes to hotel for this stop.

February 18–20, 1949
Showing at 11th St. near Union Station. Wonderful crowds here and a lot of coal smoke, too. Governor Browning and party are entertained in diner dome at close of show Sunday night the 20th. This is one of the shows advanced by Bill Voight. Fine job, Bill, and we hope you recover the spirits which were spirited from your room.

February 21, 1949
At 12:45 PM we are on the Nashville, Chattanooga and St. Louis tracks and starting for Chattanooga. A fine roadbed and scenery makes the trip all too short as we stop for the press at Cowan 3:30 PM. Very pretty trip into Chattanooga. Lookout Mountain on our right as we move into city at 5:00 PM.

February 22–24, 1949
Exhibit at NC&STL Station. Outside track near busy street with good view of train helps the attendance. Train shed roof keeps crowd dry on the one rainy day.

February 25, 1949
Staff sleeps aboard and we leave on the Southern at 3:00 AM for Johnson City and Kingsport. Two groups on this press ride. Arrive in Johnson City at 9:00 AM. Here we are turned over to the Clinchfield Railroad. At 1:00 PM Johnson City press boards and we are bound for Kingsport 20 miles away. Here the Kingsport press group joins us for the ride to St. Paul, Va., and return. Johnson City group returns from Kingsport by bus. Mr. C. D. Moss, General Manager of the Clinchfield, describes the highlights of the scenery over the P.A. as we wind around the mountain and through the tunnels.

February 26, 1949
One-day show at Union Station, Kingsport. Lovely weather, good crowd. Junior Chamber of Commerce helps with hosting the train. After close of show we pack up and leave at 11:00 PM for Johnson City, an hour's ride. For the past two days we have been host to Mr. B. K. Wingerter of Electro-Motive Division, New York, who has been so cooperative in arranging our schedules with the various railroads over which we move and on which we show.

February 27, 1949
A full day of rain as we exhibit near the ball park at Johnson City.

February 28, 1949
10:00 AM is leaving time from the site at Johnson City. Again we are on the Clinchfield and assured we have a treat in store for us on this trip to Spartanburg, S.C. A layover of an hour and a half at Erwin, Tenn., and we proceed into the Tennessee hills. As various points of interest and beauty are drawn to our attention we recall our Western trip through the Rockies. This, too, is a very scenic part of America which far too many people fail to visit. Mount Mitchell, 7,600 ft. and picturesque Table Rock are pointed out by Mr. C. D. Moss of the Clinchfield. At 3:30 PM we arrive at Marion where we meet the Spartanburg press group. This is the largest group we have had for some months. As we roll down out of the

mountains and into the flat cotton country nearing Spartanburg we begin to wonder if some of our guests are not suffering from "Bourbonic Plague." Arrive at 5:00 PM, intact.

March 1–2, 1949
Exhibit on C&WC tracks near Main Street. Good attendance helped by schools.

March 3, 1949
12:30 PM is leaving time for our last ride on the Clinchfield. Arrive at Bostic, N.C., to be switched to Seaboard Air Line tracks for remainder of trip to Charlotte. Press pick up at Ellenboro 3:15 with arrival time in Charlotte 5:15 PM.

March 4–7, 1949
Exhibit SAL freight tracks between 5th & 6th Streets. C. W. Perkins and Max Wilson busy replacing coil in steam generator.

March 8, 1949
Free day—picnic at home of Mr. Geo. Crisler, GMAC Manager, in these parts. Southern hospitality at its best.

March 9, 1949
Farewell to Charlotte at 7:30 AM as we leave SAL with a half dozen guests aboard for breakfast. Guests depart at Hamlet at 9:30 and at 10:15. We pick up our press group at famous Southern Pines. Arrive Raleigh at 11:30 AM. Special showing at 6:00 PM.

March 10–13, 1949
Exhibit at SAL Station, Raleigh, N.C. Good site and lovely weather.

March 14, 1949
Depart from Raleigh at 12:00 noon with group of GMAC men who enjoy lunch and leave train at Norlina 1:30 PM. Press pick up at Boykins 3:30 PM and arrive in Portsmouth 5:15 PM. Clearance problems here necessitates choice of another exhibition track. Still with SAL Railroad.

March 15–17, 1949
Exhibit at SAL Station. Good crowds, but on last evening of showing the line at closing time is in excess of a block long. Quick teardown necessary for 11:10 departure. All of our stops so far in 1949 have been in the territory of "Uncle" Phil Rozelle. You were a most gracious host, Phil, and we shall miss these pleasant winters in the South.

March 18, 1949
Leave Portsmouth at 11:10 PM (17th) via SAL for Richmond. At 4:00 AM we are switched to C&O tracks and ready for 8:30 departure for Charleston. Rainy and cold, but the trip is through some beautiful country and places such as White Sulphur Springs, Staunton, and Charlottesville. R. R. Seward shows his movie collection of our western trip while we travel through country almost as picturesque. Press pick up at Cotton Hill, W.Va., 4:00 PM and arrive in Charleston 5:00 PM.

March 19–21, 1949
Exhibit near C&O Station, Charleston. Poor site but fine attendance. Most outstanding event is Joe Schneiderbauer's four-day holiday in Chicago. Lee Gentry puts his social activities into high gear at this point.

March 22, 1949
Leave Charleston with Huntington press aboard at 10:30 AM for the hour's ride to Huntington, arriving 11:30 AM. Show hours at C&O Station start today at 3:00 PM

March 23–24, 1949
Exhibit at C&O Station, Huntington. We were told this town would be very railroad conscious, and it really is.

March 25, 1949
Depart from Huntington at 2:00 PM for Lexington, Ky. Stop made at Morehead, Ky., to pick up press group at 3:15 PM with a 5:00 PM arrival in Lexington.

March 26–28, 1949
Exhibit at C&O Station, Lexington, Ky. Feature of this stop for the train staff is a visit at Calumet farms, arranged by Robt. Emerick, P.R. Regional Manager. Thanks, Bob, it was a welcome relief from the smell of railroad smoke.

March 29, 1949
Leave Lexington at 4:30 AM via C&O to Huntington where we are switched to Baltimore and Ohio tracks at 9:30 AM. Depart from here at 10:45 and follow B&O line to St. Mary's, where press boards at 3:05 PM and accompany us to Wheeling, arriving 5:15 PM. Rough track makes this a rather slow movement.

March 30–April 1, 1949
Exhibit at B&O Station on elevated tracks. Train staff entertained by Ellsworth E. Seitz, former train staff member.

April 2, 1949
Move toward Elmira started at 1:00 AM on B&O on towards Pittsburgh, thence on mail line of Pennsylvania at 4:00 AM, arriving at Altoona 6:50 AM. Fuel here and leave at 7:15 via Bald Eagle branch line of PRR for Lock Haven arriving 8:45 AM. Train wyed and head north for Olean, N.Y., where at 1:08 PM we are switched to the little used freight line and pointed toward Mt. Morris. Schedule calls for 3:00 PM arrival, but ten-mile speed and rough road brings us to Mt. Morris at 4:35 PM. We are quickly switched to Delaware, Lackawanna & Western Road, backed eight miles to wye at Dansville and pull up at station at exactly 5:00 PM, one and one-half hours late. Press group still in good humor and enjoy 80 mile an hour ride into Elmira, arriving at 6:15 PM.

April 3–4, 1949
Exhibit at DL&W freight yards, 5th & State Streets. Poor site but good attendance. Train moved out each nite.

April 5, 1949
Leave Elmira at 10:00 AM with small press group for Binghamton who arrive by bus. Arrive Binghamton 11:30 AM. Passengers discharged and train is wyed and returned to site in freight yards. Show opens 3:00 PM with acute shortage of hosts.

April 6–7, 1949
Exhibit at DL&W freight yards. Rain part of each day.

April 8, 1949
Press group arrives via train at 3:05 PM and at 3:30 PM we are loaded and headed for Scranton via scenic DL&W. Large group of 179 press people. Arrive Scranton at 5:00 PM.

April 9–10, 1949
Exhibit at Scranton, lower DL&W tracks at foot of Washington.

April 11, 1949
Leave Scranton 9:00 AM on DL&W for switching to New York, Ontario & Western at 10:30 AM and proceed toward Poughkeepsie. Flowers ordered enroute and picked up at Liberty. Mr. & Mrs. B. K. Wingerter of New York are guests aboard as far as Middletown, where we leave NYO&W for the NYNH&H. Press pick up at Maybrook, 3:50 PM. Arrive in Poughkeepsie, over high bridge spanning the Hudson, at 5:00 PM.

April 12–13, 1949
Exhibit in NYNH&H freight yards at Poughkeepsie.

April 14, 1949
Leave Poughkeepsie 9:00 AM on NYNH&H for trip to New Haven, Conn. Press pick up at Danbury 10:15 AM, arriving New Haven at 11:45 AM. Show opens at 3:00 PM near passenger station.

April 15, 1949
Good Friday, and second day of New Haven exhibit.

April 16, 1949
Leave New Haven at 9:05 AM from Bridgeport via Danbury for press pick up at 10:20 AM, returning to sea coast again by different route of NYNH&H, arriving at 11:45 AM. Show opens at 3:00 PM at Fairfield & State.

April 17, 1949
Easter Sunday showing at Bridgeport. Early show plans fouled up and Joe and his crew play hosts to 800 persons.

April 18, 1949
Leave Bridgeport 9:55 AM for ride up the main line of NYNH&H to the submarine formed town of New London. Press pick up at New Haven 10:20 AM, arrival time in New London 11:30 AM. Show 3:00 PM at Fort Yard near Howard St.

April 19, 1949
Second day of New London exhibit. Harrison Daniels of Mr. Schneiderbauer's crew mails his 100th postcard from here. A wide acquaintance. Ed Kraigel proves to be a fine choice to replace Max Wilson for EMD. We suspect he has his own valet.

April 20, 1949
Press aboard as we depart from New London at 10:45 AM for Providence via NYNH&H. Two-day shows mean lots of work for the Duss and Fleming team and their crews. Arrive at Providence 11:30 AM. Show 3:00 PM.

April 21–22, 1949
First three-day show in many stops. Staff enjoys the lovely Sheraton Biltmore and take the opportunity to catch up on the laundry problem. Farewell gathering for Max Wilson, leaving for a "Stationary" job in LaGrange, Ill.

April 23, 1949
1:30 PM is departure time from Providence as we leave on the NYNH&H road for transfer to Boston & Maine road at Worcester 3:00 PM. Pick up point is Ayers, Mass., at 4:00 PM with 5:00 PM arrival in Manchester, N.H.

April 24–25, 1949
Exhibit at Manchester Station. Very cold weather. The usual confusion as the daylight time goes into effect.

April 26, 1949
Leave Manchester via B&M at 12:45 PM with lunch aboard. Press pick up at 3:05 PM at Dover and arrive in Portland, Me., at 4:20 PM.

April 27–28, 1949
Exhibit at Union Station, Portland. Mr. Ralph Moore and Mr. W. E. Hamilton of Detroit meet with staff to talk over Chicago Fair plans and future outlook.

April 29, 1949
Leave Portland at 9:00 AM for trip via B&M to Worcester. Press pick up at Lowell. Mr. Ralph Moore is with us for this trip. Arrive at Worcester 12:00 noon and after our guests detrain we are routed to the Pullman plant for short showing to employees and families at-

tending Safety Award ceremonies. Mr. Steve Early, speaker. Leave Pullman 7:00 PM and at 7:20 are on location at Grafton and Franklin St.

April 30–May 2, 1949
Exhibit at B&A freight yards, Worcester.

May 3, 1949
Leave Worcester 1:45 PM via B&A for Springfield. Press board at Worcester for hour's ride. Arrive 2:45 PM.

May 4–6, 1949
Exhibit in B&A freight yards, Springfield at Corner Springfield and Taylor.

May 7, 1949
Leave Springfield at 9:45 AM via NYNH&H for Hartford with press aboard. Arrive 10:30 AM and open for exhibit at 2:00 PM. Location, Spruce St. yards adjacent to station.

May 8–9, 1949
Exhibit at Hartford at 10:45 PM (9th) we leave via NYNH&H for Springfield, thence to B&A lines for Albany, arriving 2:00 AM. Mr. Milton Gearing of New Departure and party are aboard.

May 10, 1949
Quick switch to NYC at Albany and at 2:10 we are off for Buffalo. Fuel at Buffalo 8:30 AM, arrive in Cleveland (Penn Station) at 11:00 AM. Train is cleaned and at 1:30 PM we are switched to B&O at Whiskey Island. Pick up large press group at Southpark 2:40 PM, arriving in Canton at 4:30 PM. Train backs to Akron Jct. for turning and is back at site 8:00 PM. Republic Steel have big dinner which is attended by N.D. people.

May 11–13, 1949
Exhibit at Canton on B&O tracks at West 7th St. near park. High cinder embankment on show side makes this a poor site.

May 14, 1949
Pull out of Canton on B&O in reverse to Sandyville at 7:30 AM. Press boards here at 9:00 AM for ride to Akron arriving 10:30 AM. Poor show due to B&O showing of Columbian two days before.

May 15–16, 1949
Exhibit at Howard Street Station, Akron. Bill Hammond and Bill Kirk get in some target practice in the "wilds of Akron."

May 17, 1949
Early morning departure (5:50 AM) via B&O for Lima, arriving 9:00 AM. One-day showing only, near Penn Station. Mr. W. E. Hamilton is luncheon speaker at Country Club.

May 18, 1949
Group of Lima guests board at 2:35 PM for ride to Defiance via B&O, arriving 3:45 PM. Here we are switched to Wabash road and at 4:20 PM we are rolling toward Ft. Wayne with new group of Ft. Wayne people. Arrive at 4:35 PM (CST).

May 19–20, 1949
Exhibit at Wabash freight yards on Harrison St. Much rain for Ft. Wayne show.

May 21, 1949
Train crew sleeps aboard on night of 20th and 3:00 AM is departure time for Defiance via Wabash. Arrive 5:50 AM and at 7:30 AM we are "Anchored" at GM Plant outside of city. B&O does switching. After much rain, plant grounds are very soft. Open house showing. Defiance needs a good hotel.

May 22, 1949
Second day of open house showing. Defiance turned out in large numbers.

May 23, 1949
Leave plant at 8:05 AM via B&O for Auburn where we switch to NYC. Reverse move to Waterloo and forward to Kendallville for press pick up. Group is one hour late due to detour. Leave Kendallville at 11:30 AM arriving South Bend at 11:35 (CST). Showing at 3:00 PM.

May 24–25, 1949
Exhibit at South Bend, on upper level of Union Station.

May 26, 1949
Leave South Bend at 7:00 AM via NYC for Englewood, arriving 8:30 AM. Switch to 51st Street yards for Pullman cleaning and then to Dearborn Station. Leave at 11:00 AM with small group of C&EI people on C&EI road. Group detrains at Danville 1:25 PM and at 3:35 Evansville press boards at Vincennes. Arrive Evansville 5:00 PM.

May 27–28, 1949
Exhibit at C&EI freight yards, Evansville.

May 29, 1949
6:00 AM departure from Evansville on C&EI to Dansville. Switch to Wabash and press from Decatur boards at 10:30 AM. Arrive 12:00 noon. Show 3:00 PM, close to "hot track" in depot.

May 30, 1949
Second day of Decatur show. Small group of train staff journey to Indianapolis to view the Speedway Classic.

May 31, 1949
Press from Springfield board as we leave for Springfield at 1:55 PM via IC. Arrive in Springfield at 3:15 PM. One guest, Mr. Stubbs, publisher, has bad heart attack, but recovers sufficiently to be taken to his home. Mr. C. R. Osborn is honored at ceremony on platform in front of locomotive. Plaque presented by State Senator.

June 1–2, 1949
Exhibit at IC freight yards, Springfield.

June 3, 1949
9:00 AM is departure time for Peoria. Press are aboard as we leave. Much time consumed because of switch from IC to Pekin Union R.R. 10 miles out of Peoria. Arrive 11:45 AM. Showing 2:00 PM on CRI&P tracks near river.

June 4–5, 1949
Exhibit at beautiful site near river, Peoria.

June 6, 1949
Mr. Ralph Moore and Mr. J. D. Farrington each with small group are luncheon guests as we leave Peoria at 11:00 AM via Rock Island lines. Arrive Rock Island City 1:30 PM. Show at 4:30 PM. At 10:00 PM we move across river to Davenport site.

June 7–8, 1949
Last city of the tour, Davenport, which we had to omit last year because of threatened rail tie up. Showing at 5th and Iowa. Local hosts were especially cooperative.

June 9, 1949
Leave Davenport 6:00 AM via CB&Q lines for St. Louis. Stop at Rock Island for small group of dealers and hosts who ride to Galesburg. By 3:00 PM. we are in Pullman Yards, St. Louis, for limited overhaul prior to Chicago Railroad Fair. Mr. J. F. Fitzgerald and two sons are with us for this run as far as Bureau.

June 10–20, 1949
Shopping period at St. Louis Shops.

June 21, 1949
Everyone aboard by 5:00 PM. At 8:00 PM we move to Union Station, and by 11:30 PM are on our way to Chicago via IC. 6:30 AM is arrival time at IC Station, Chicago.

June 22, 1949
Train is cleaned in IC yards for run to Gillman and return with press group and IC officials. Leave Chicago at 1:00 PM and return at 3:55 PM. 100 mile hour speed maintained most of the way. After guests detrain, we move to Fairground site and are placed at 6:00 PM.

June 23–24, 1949
Final preparations made for fair opening. This year's Fair showing is to be better in many ways. We have a full pavement twelve feet from rail to outside stanchions, stationary stanchions with 9 stationary floodlights. Site is close to grandstand and "backstage" activity of pageant "Wheels a-Rolling." Arrangements have been made for crew of 36 college boys to host train for entire Fair period. Mr. Seward and Mr. Gentry have done a fine job in scheduling hours for these boys and a system of position rotation is worked out. Mr. Hamilton and Mr. Fitzgerald deserve credit for their selection of a fine group of boys.

June 25, 1949–September 16, 1949
Fair opens 10:00 AM. Crowds are somewhat smaller than last year but continue to grow as Fair progresses. More attractions, including the Ice Show, Water Carnival, Gold Gulch, etc. Eating accommodations are improved with C&O, IC, and CRI&P diners serving to the public. Also at the Fair is San Francisco's famed cable car, and the giant statue of Paul Bunyan attracts large crowds. By mid-August our attendance has grown 10,000 per day, weekends 12,000. As was true last year, the pageant is a great attraction and worthy of fine comment. A bigger and faster show even surpassing last year's production. Various small groups of GM executives and others are dinner and luncheon guests aboard during the Fair showing.

Canadian Tour
September 17–20, 1949
After close of the show on the 16th, the ramps, etc. are torn down, the equipment on the track behind us was moved out and at 1:10 AM we move out of Fair Grounds for the last time. On board are a few of the boys who have been our hosting staff for the summer. Proceed to 103rd St. Plant for reconditioning preparatory to leaving on our Canadian tour.

September 21, 1949
Leave 103rd St. Plant at 2:00 AM for delivery to Dearborn station via Pullman Railroad and Indiana Harbor Belt line. Staff boards at Dearborn Station 7:00 AM for trip to Detroit. As guests we have nine boys of the host crew plus Mrs. J. G. Schneiderbauer and Miss Lucie Schneiderbauer. Boys detrain at South Bend, the ladies accompanying us to Detroit. Arrive via Grand Trunk at 2:00 PM Milwaukee Jct. Transferred to Michigan Central coach yards for cleaning.

September 22, 1949
Staff sleeps aboard on night of 21st. Early move to Windsor thru M.C. tunnel. Leaving time from CNR station in Windsor is 9:15 AM. Press pick up at Chatham 10:30 AM. Arrive London 12:15 PM. Luncheon is served to guests as soon as train is spotted at CPR Station. Mr. E. V. Rippingille Jr., President of General Motors Diesel of Canada, is speaker. Group includes London civic and business leaders. While luncheon program is in progress, train is made ready for 4:00 PM showing to ticket holders. Shortly after opening one of the very

few mishaps to occur on the entire tour takes place. Hand brakes on each car had not been set as had been the custom and when Engineer's Instructor, Ed Kraegel, released the air on the locomotive, the train rolled slowly backward about 12 feet, dragging the front ramp and pushing rear ramp along. Luckily, only one person was on the rear entrance ramp at the time and outside of a jar when he jumped to the ground, was unhurt. Canadian Pacific men arranged for a quick welding job on rear ramp and by using observation car vestibule for entrance during this period, we continued business as usual. Our U.S. tour started off with a slight mishap, too, on May 27, 1947, so this may be a good omen.

September 23–24, 1949
Exhibit near CPR station, London. Long lines would indicate that the T.O.T. still had crowd appeal in Canada as well as elsewhere.

September 25, 1949
Leave London at 7:30 AM for our first run on Canadian Pacific rails, for Ottawa. Breakfast and lunch aboard. Guests include Mr. W. A. Wecker, Mr. A. E. McGilvray, and Mr. T. R. Elliott of GM of Canada. Press pick up at Smiths Falls, 3:30 PM. Arrive Ottawa 5:00 PM. Chateau Laurier across from station is our first CNR hotel.

September 26–28, 1949
Much rain for this show. Exhibit near station. On 26th, cabinet members are entertained by Mr. C. E. Wilson at a luncheon while riding out to Carlton Place and back. Left 11:30 and returned at 1:30 PM. This trip also on CPR.

September 29, 1949
1:00 PM is departure time from Ottawa. Lunch aboard as we roll towards Montreal via CNR. Press pick up at Coteau 3:30 PM arriving in Montreal at 4:45 PM.

September 30, October 1–4, 1949
Canadian National went all out to make our show here at Montreal a success. Fence was created to separate T.O.T. from main line tracks and very necessary, too, considering that here we handled enormous crowds. On Sunday October 2nd at 4:00 PM there were by actual count 3,619 people waiting in line to visit train. Exhibit site was two blocks from CNR terminal, at Inspector St. It was here that Cy Perkins received word of his son's very serious injury in an auto accident and left us for his home in Newton, Kansas.

October 5, 1949
This trip again in CPR lines. At 9:00 AM we board T.O.T. at site and after much switching are left at North Montreal Station to await a small party of guests. At 11:30 AM we depart for Quebec. Press pick up at Trois Rivers at 3:15 PM of group arriving by train from Quebec City. Approach Quebec in the gathering dusk, arriving at CPR station in the "Old City" at 4:30 PM. Staff quartered at the famous Chateau Frontenac of the CPR chain. Mr. Williams Davidson is speaker at the Winter Club luncheon. CPR police are especially cooperative and efficient. Governor General and his family visit us at noon on the 7th.

October 6–8, 1949
Exhibit at CPR station at bottom of hill leading down from the Chateau which dominates city. French interpreter engaged to handle Public Address system for this showing. Crowd is 90% French speaking. Weather warm and sunny by day and very cold at night. Local GM people most generous with time and cars in showing us their community.

October 9, 1949
Leave Quebec on Quebec Central road at 6:00 AM in heavy fog for the run to Sherbrooke in the southern part of Quebec Province. Breakfast aboard as we pass over the big bridge at edge of city. Several stops made to pick up small groups for short rides between points. Press pick up at Thedford Mines, the asbestos center of the country. Arrive at Sherbrooke

11:30 AM. Beautiful warm day. Show opens 2:00 PM. Train staff spread around among five different hotels. Mr. Paule LeProhon, Frigidaire dealer, is host to press group and train staff at luncheon.

October 10, 1949
Second day of Sherbrooke showing at site near CNR station. This is Canadian Thanksgiving Day and crowd is large. Many special groups visit before opening at 2:00 PM. Arno Schneiderbauer (Little Hat) who is replacing Don MacShane for this part of tour gets an expensive French lesson concerning truck rentals.

6:30 AM is departure time as we leave CPR station for run over their lines to St. John on the outskirts of Montreal. This is our last trip over CPR and at 9:15 AM we are on CNR road for balance of trip to Oshawa, home of GM of Canada. Group of 20 picked up at Cromwell for short ride to Brockville. Luncheon served as we leave Brockville. Press pick up at 3:20 PM at Cobourg, arriving Oshawa 4:30 PM. Special showing for GM supervisors and families at 7:00 PM. Train site is adjacent to main line. Light standards must be anchored to prevent falling from jar of passing trains.

October 12–13, 1949
Exhibit at CNR station, Oshawa. 12th restricted to ticket holders only. Here we have new rubber collar replaced in one of the Hyatt bearings with the assistance of Hyatt man rushed from Harrison, N.J.

October 14, 1949
Leave Oshawa on CNR at 10:45 AM for short trip to Toronto. Press aboard as we leave. Arrive Toronto 11:45 AM and are spotted at old North Toronto station, no longer in use. Fine exhibition site. Train stands on overpass in sight of all Yonge St. traffic.

October 15–19, 1949
Exhibit at North Toronto station. While this is our second visit to this city (Sept. 1947) our crowds were excellent. Splendid cooperation from local GM people on hosting problem. Mr. Cannon of Frigidaire entertained staff at his home on Sunday, 16th. The Schneiderbauers, both Jr. and Sr., also do something entertaining. Johnny Duss takes his first plane ride home to Pittsburgh for the weekend. Frigidaire luncheon aboard train on 18th. This Canadian trip, being financed by the Canadian Divisions, gives a breather to Miss Dorothy Scott and Miss Betty Neff of the Business Management Section.

October 20, 1949
Again on CNR as we leave Toronto at 10:45 AM for short run to Hamilton. Mrs. Joseph Malloy, whose husband has been with us as business representative for CNR, is among guests for this ride, as are several people from MacLaren Advertising Co. who have handled publicity for this tour. Arrive in Hamilton after back up move into city at 11:45. Showing 3:00 PM. Mr. Roy Hallem of Frigidaire again on the job with a full crew of hosts. As train can be seen from two different overpasses, the crowd come in from all directions, making handling very difficult. Error in advertising posters as to location of train site adds to confusion.

October 21–22, 1949
Exhibit at CNR Station, Hamilton. Lovely weather and fine crowd.

October 23, 1949
Sunday. 4:00 PM departure from Hamilton via CNR with St. Catharines press aboard. Group not identified with badges makes loading a problem. A short run to St. Catharines, but after turning train at the wye in Welland it is 6:00 PM before we arrive on site at St. Catharines.

October 24–25, 1949
Exhibit at CNR station, St. Catharines. Cold weather makes it necessary to procure steam engine from Niagara Falls for heating. It was here that we had an unruly crowd in September 1947, but perhaps that generation has grown up, for all is calm and peaceful for this showing.

October 26, 1949
Breakfast aboard as we leave St. Catharines at 8:00 AM and proceed up the CNR for Stratford and one-day show. Weather cold and bright. Press pick up at Paris 10:00 AM with 11:00 AM arrival at Stratford. Another steam engine for heating. Bad crowd situation as line crossed hot track, but utmost caution used by passing trains.

October 27, 1949
8:00 AM sees us on our way to Chatham via CNR. Stop made in London for press group and also group of ladies as guests of Mrs. E. V. Rippingille Jr. Arrival in Chatham 11:00 AM for one-day showing starting at 2:00 PM. Site is one block from CNR station.

October 28, 1949
Leave Chatham 10:00 AM on our last press ride. Large group of Detroit and Windsor GM people with press group. Mothers of staff members Lee Gentry and Wm. Hammond also in group. Arrive in Windsor 11:00 AM prior to our last show at CNR station, Windsor. Staff dinner tonite with Miss Betty Bell as honored guest. As secretary to Mr. W. E. Hamilton, she has been in the thick of the T.O.T. program. Thanks, Miss Bell, for the many courtesies shown the entire train staff.

October 30, 1949
Transfer to Essex Terminal Railway at 12:00 noon and then to NYC for trip through the tunnel to Detroit. After still another transfer to Grand Trunk at Milwaukee Jct. we leave Detroit at 4:00 PM. Our last ride on the Blue Lady. The T.O.T. has been home to most of us for two and one-half years. Looking back on these many months of close association with other staff members makes one pause and silently salute each and every one. They were gentlemen, all. Arrive at the last port of call, Dearborn Station, 8:30 PM. T.O.T. is moved to 103rd St. Plant for cleaning and overhaul, to be eventually sold into railroad service. So long, Blue Lady, for us you will always be the Train Of Tomorrow.

APPENDIX B

TECHNICAL INFORMATION

Appendix B is devoted to details about the mechanical systems, furnishings, and supplies of the *Train Of Tomorrow*. This includes some of the innovations unique to the train, reprinted with written permission of Pullman Technologies, Libbey Glass, Union Pacific, and the General Motors Corporation.

Air-Conditioning and Ventilation

All four cars are fully air-conditioned with Frigidaire equipment. Each car has a 10-ton capacity, split into 4-ton and 6-ton units. The units are mounted overhead in the car's

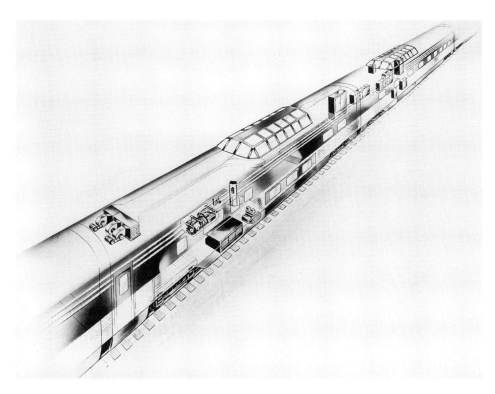

General Motors railroad product placement on the coach and dining car of the *Train Of Tomorrow*. GM.

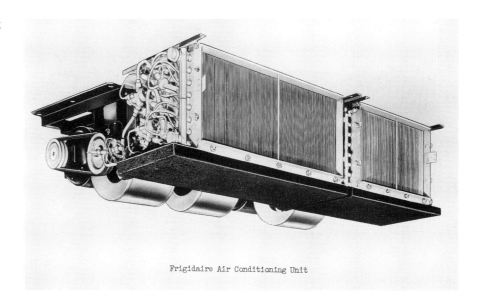

The Frigidaire air-conditioning unit. GM.

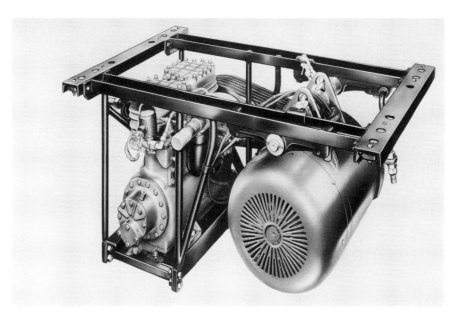

The Frigidaire air-conditioning compressor unit. GM.

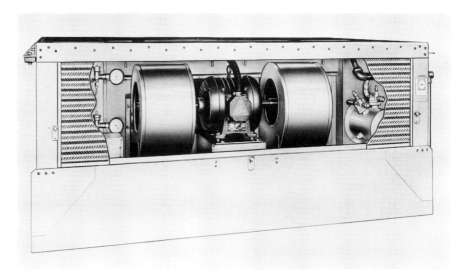

The Frigidaire air-conditioning condenser unit. GM.

TECHNICAL INFORMATION

The air-conditioning condenser and compressor units, mounted under one of the cars. UPRR.

body. Blower fan capacity for the chair car, dining car, and observation car is 2,800 cubic feet per minute (CFM). The sleeping car blower fan has a capacity of 2,400 CFM with 25 percent of the air in the sleeping car being fresh. Each dome compartment is supplied with 1,000 CFM of conditioned air. The remainder of the air is directed to the lower level rooms.

Batteries

The train is equipped with a set of 32-volt batteries for emergency lighting and for starting the diesel alternator power packages. A 6-volt automobile type battery is used on the observation car to power the radiotelephone system.

Receptacles are provided on both sides of each car to charge the 32-volt batteries. Two additional receptacles, one on each side of the car, are provided on the dining car to charge the 32-volt battery for the 40 kilowatt (kW) power unit. A belt-driven exciter mounted on the diesel engine of the power units also handles battery charging.

Two battery boxes are provided on each car to house the batteries needed to start the diesel engine. An extra battery box has been added to the dining car for starting the diesel engine in that car. Another extra battery box has been added to the observation car to house an automobile type battery to provide power for the radiotelephone system.

Berths and Beds

The mattresses are 4½" thick and of spring and rubber construction, except for the beds in the lower roomettes. These are made of two sheets of cored rubber with interlaced hair between, not exceeding 3½" in thickness.

The drawing rooms have two upper berths and one convertible sofa that converts to a lower berth. All the berths are mounted lengthwise in the car. One upper berth folds down from the wall, and the other one drops down from the ceiling. The mattresses on the upper berths measure 32" × 73", while the mattress on the lower berth is 35" × 75".

The compartments are equipped with a sofa that converts to a lower berth and another lower berth that folds down out of the wall. The two berths are placed lengthwise in the car and are outfitted with mattresses measuring 32" × 75".

The berths in the upper roomettes fold out of the wall and have mattresses that are 32" × 74". The sliding berth in the lower roomette has a mattress that measures 32" × 72".

Brakes

The Westinghouse HSC air brake system has "individual speed governors and electro-pneumatic straight air control."[1] An electric trainline is provided for the electric air brakes. "A receptacle is located at the end of all of the cars and one complete intercar connector is provided for each car."[2]

Each car has a hand brake located at only one end of the car. For cars where the brake control is located in the vestibule, a wheel handle–type hand brake is used. A pump handle is mounted on cars that have the brake on the dummy end. Under AAR guidelines, the hand brakes need to operate only one truck per car.

Car Supplies

General Motors is responsible for applying some of the supplies and materials used on the cars. For all cars it includes the following:

- Linen towels
- Paper towels
- Toilet paper
- Cake soap
- Bags for soiled linens
- Book matches
- All cleaning materials: bags, soaps, sweepers, etc.
- Crew instruction cards
- Living plants

For the chair car, *Star Dust*, the following items are added:

- Paper drinking cups
- Sanitary napkins
- Pillows and pillowcases
- Porter's coat

For *Sky View*, General Motors is responsible for supplying the following items:

- Table linen: table cloths and napkins
- Silverware: knives, forks, spoons, etc.
- Silverware: dishes, coffee pots, creamers, etc.
- Chinaware: dishes
- Glassware including water bottles
- Trays: serving and cash
- Menus, pencils, checks, etc.
- Ashtrays
- Seat covers
- Waiters' and cooks' uniforms
- Kitchen and pantry loose equipment
    - Pots, pans, skillets, etc.

TECHNICAL INFORMATION

    Culinary equipment including utensils, such as knives, forks, ladles, pie tins, cake tins, etc.
    Mixing bowls
    Canisters for coffee, tea, sugar, spices, etc.
    Pans and jars for hot tables
    Ice cream cans
    Ice cream scoops
    Dish towels and glass towels
    Cleaning compound for dishes
    Spare brushes for glass washer
    Spare racks for dishwasher
    Drink mixer

Dream Cloud has the following materials and supplies requirements:

    Linen for the berths
    Sheets and pillowcases
    Blankets
    Paper drinking cups
    Sanitary napkins in the general toilet
    Porter's coat
    Hotel Red Book, timetables, telegram blanks

The sleeping car porter has the following supplies to provide efficient customer service:

    Shoe polish and brush, clothes brush, hat brush, paper hat bags, and whisk broom
    Formaldehyde for deodorizer jug
    Salt baggage
    Quiet sign
    Fly swatter
    Baby guards
    Ironing boards, sleeve board, pressing iron and stand

Materials and supplies for the observation lounge car, *Moon Glow*, include:

    Radiotelephone equipment including telephones
    Telephone at the observation end
    Writing pens, ink, and stationery at the desk
    Ironing board, sleeve board, pressing iron and stand
    Blotters for the desk with corners
    Broom: toy size no. 3
    Dust pan
    Fly swatter
    Electric drink mixer, juice extractor, and hot cup at the bar
    Uniforms for the bartenders and waiters
    Table linens and napkins
    Liquor license
    Sanitary napkins in the women's toilet
    Loose bar equipment
        Glassware
        Swizzle sticks and muddlers
        Trays for serving and cash
        Jars at the bar.
        Small bar tools, such as bottle openers, can openers, mixing spoons
    Wire recorder spools (two furnished with the set)[3]

Ceilings

Ceilings throughout the train are covered with aluminum, except in the kitchen, pantry, and bar, where stainless steel is used.

Center Sills

A center sill is a beam or group of beams that supports the major weight of the car. "Between bolsters up to the lower level under the Astra Dome consists of two AAR Z-26 sections, 31.3 pounds per foot, with the top flanges welded together continuously, complete section providing a total area of 18.4 square inches. The center sill is welded behind each bolster to a separately constructed built-up draft sill of welded construction. At the lower level under the Astra Dome, the Z-26 center sill stops and is attached to heavy beams placed crosswise of the car. These beams are connected at the ends at the side post lines to the box section members, which serve as the center sill for this portion of the car."[4]

China, Glassware, and Silver/Flatware

The *Train Of Tomorrow* was expected to be on tour for only six months when it left Chicago on its first eastern tour. But the dining car, *Sky View*, was fully equipped with china, silverware, flatware, and glassware, enabling guests of General Motors to enjoy meals being served in the three dining rooms, including the one in the dome, during various public relations functions. The International Silver Company supplied the silverware and flatware. The china was made by the Onondaga Pottery Company, now the Syracuse China Company, and distributed by E. A. Hinrichs of Chicago. Libbey glassware was used not only in the dining car but in the observation lounge car.[5]

Ed Pohlman, manager of advertising and public relations, identified the glassware by Libbey as a stock pattern named Georgian, which came in 23 pieces:

Cordial—1 oz.
Champagne—4½ oz.
Champagne—5½ oz.
Cocktail—3½ oz.—shallow design
Cocktail—3½ oz.—deep design
Cocktail—4 oz.—deep design
Cocktail—4½ oz.—deep design
Lined Cocktail—4½ oz.—deep design
Wine—3 oz.
Rhine Wine—4 oz.
Claret—4 oz.
Sherry—2 oz.
Whiskey Sour—4¼ oz.
Parfait—4¼ oz.
Irish Coffee—6 oz.
Irish Coffee—7 oz.
Footed Rocks—5 oz.
Footed Rocks—7 oz.
Footed Hi-ball—10 oz.
Low Goblet—11½ oz.
Goblet—8 oz.
Goblet—9 oz.
Pilsner—10 oz.

TECHNICAL INFORMATION

From the left: 2 oz. cordial, 1 oz. cordial, 8 oz. goblet, 11½ oz. low goblet, and 9 oz. goblet. Libbey Glass, Inc.

It is not known exactly how many pieces were on the *Train Of Tomorrow*, but photographs of the dining car and observation lounge car show the 8 oz. and 9 oz. glasses (used for water and milk), the 2 oz. sherry glass, the 4 oz. claret, the 1 oz. cordial, and the 3 oz. deep design cocktail glass.

Photographs show drinks being served in the observation lounge car in 8 oz. hi-ball glasses from a pattern called Straight Sided.

Richard Luckin, author of *Dining on Rails: An Encyclopedia of Railroad China* and *Teapot Treasury*, identified the Onondaga china as a custom design called the *Train Of Tomorrow* pattern. All of the pieces had a white background. The plates, saucers, and bowls had a wide band of light green just below the rim with a pinstripe of gold on the outside edge. In the center was a large white rose with dark green leaves. The teapots,

From the left: 4½ oz. cocktail, 4½ oz. lined cocktail, 4½ oz. champagne, 3½ oz. cocktail, 5½ oz. champagne, 4 oz. cocktail, and 3½ oz. cocktail. Libbey Glass, Inc.

A small bowl from the *Train Of Tomorrow* collection. Richard Luckin Collection—All rights reserved.

cocoa pots, cups, and soup bowls had a gold pinstripe just below the rim, with two of the big white flowers with dark green leaves on each side of the piece. There was a gold pinstripe on the handle of each piece as well as on the spout of the teapot and cocoa pot. The lids on the teapots and cocoa pots were the same light green as the rim of the plates with a pinstripe around the edge of the lid. There was also a gold pinstripe around the top of the pot rim. The manufacturer's name for the teapot was Lobe, while the manufacturer's name for the 11-ounce cocoa pot design was Stanley, and it was a very popular design used on many railroads such as the Santa Fe, New York Central, and the Gulf Mobile and Ohio in the late 1940s and early 1950s.

A small plate in the *Train Of Tomorrow* pattern. Richard Luckin Collection—All rights reserved.

TECHNICAL INFORMATION 185

A coffee cup from the *Train Of Tomorrow*. Richard Luckin Collection—All rights reserved.

The china collection is believed to have included the following items:

    Dinner plates
    Small plates
    Salad and cereal bowls
    Side dishes (Monkey dishes)
    Coffee cups
    Cup saucers
    Double-handled soup cups
    Soup cup saucers
    Teapots
    Cocoa pots

The salad fork, dinner fork, dinner knife, soup spoon, and teaspoon. Rudy Morgenfruh Collection—All rights reserved.

The flatware, made by International Silver Company, now known as Insilco, was identified by Rudy Morgenfruh, an authority on railroad silverware and flatware, as the pattern called Century B.[6] Originally produced for use on the *20th Century Limited* of the New York Central Railroad, the pattern was eventually used by 10 other railroads.

Morgenfruh speculates that because the hollowware was also purchased from International Silver Company, it was probably an open-stock pattern, since it did not match any pattern used by any railroad or Pullman. However, Morgenfruh did note that the sugar bowls, water carafes, and table linens were probably provided by Pullman. Several of the tablecloths that were photographed in the dining car have the paired maple leaf as well as the name Pullman visible in the pattern.

Coach Seats

High-back coach seats are used in the coach section of the chair car, while low-back coach seats are used in the Astra Dome of the chair car, the sleeping car, and the observation car.

All seat backs recline (the coach seats recline to nine positions), and the entire unit rotates, so the seats can be turned to face the direction the car will be traveling or to make facing seating arrangements for groups who are traveling together. Each seat is equipped with a footrest and an ashtray in the armrest.

A Heywood-Wakefield "Sleepy Hollow" coach seat. UPRR.

TECHNICAL INFORMATION

Clearances

The cars will clear on a 23 degree curve with a 250.8 feet radius. The lower portion of the cars clears third rail clearances.

Couplers

The couplers are AAR Type H Tight Lock Couplers, applied on the ends of all the cars, except the observation end, which has a small Traxco-type coupler. This coupler is made to swing in toward the car and be covered by skirting when not in use. In addition, a spare Type E coupler is supplied for future use on the observation end.

On the Type H couplers, Type no. 6 uncoupler mechanisms are located on both sides of each car. Uncoupling mechanisms for both the Traxco-type and Type E couplers are on both side of the observation end.

The couplers of the *Train Of Tomorrow*. UPRR.

Diaphragms

All the car ends, except the observation end, are supplied with Pullman-Standard inner and outer diaphragms. The inner diaphragm is a twofold type made of three-ply canvas belting, treated with a fire-resistant, waterproof compound. The outer diaphragm is made of rubber.

Dimensions (rounded to the nearest inch)

Length of the train	411' 0"
Length of the cars over the coupler pulling faces	85' 0"
Length between truck centers	59' 6"
Width, over the side sills	10' 0"
Width, inside between the posts	9' 6"
Height, rail to top of roof carlines at ends (nominal)	13' 6"
Height, rail to top of Astra Dome carlines (nominal)	15' 6"
Height, rail to top of platform floor plate	4' 2"
Height, rail to top of floor at ends	4' 3"
Height, rail to top of floor, lower level rooms	3' 0"
Height, rail to top of floor, main passageway	3' 2"
Height, rail to top of floor, Astra Dome passageway	8' 8"
Height, rail to center line of coupler	2' 11"
Width of vestibule platform over top step riser	4' 5"

Depressed Floor under the Astra Dome

The depressed floor under the Astra Dome "consists of four crossbearers, one located at each end of the depressed portion of the car with two equally spaced between. Crossbearers extend across the width of the car horizontally. At the sides of the car they are fitted and welded to built-up I-sections placed longitudinally in the car. These longitudinal members extend continuously the full length of the depressed portion."[7]

Electrical Equipment

Electrical power for each car is provided by one 3-cylinder, 1,200 rpm diesel engine generating unit, built by the Detroit Diesel Division, and is mounted underneath the car. "The key is a 'split-alternator'—simply described as two alternating current generators built on a common shaft and in a common frame."[8] Alternator no. 1 generates 10 kW and supplies 120-volt, 60-cycle, three-phase power for lighting, small appliances, and evaporator blower motors. Alternator no. 2 generates 15 kW of power and provides 220-volt, 60-cycle, three-phase power for the air-conditioning system.

Each unit is mounted on a slide-out track for ease of maintenance and repair and is tilted on its crankshaft axis to provide better roadbed clearance and to give better access to the valve rockers, fuel injectors, oil filler tube, governor, and other parts.[9] A steam heat coil is mounted in the alternator's enclosure to prevent the temperature from falling below 50 degrees in the compartment during winter. A Frigidaire radiator unit is mounted underneath the car, adjacent to the diesel-alternator unit. Each car is also equipped with a 100-gallon fuel tank with an emergency fuel shutoff on both the interior and the exterior of the car. Each tank has a level gauge mounted underneath the car and a liquidometer with the dial calibrated in gallons mounted in the interior of the car. "It is said that use of

TECHNICAL INFORMATION 189

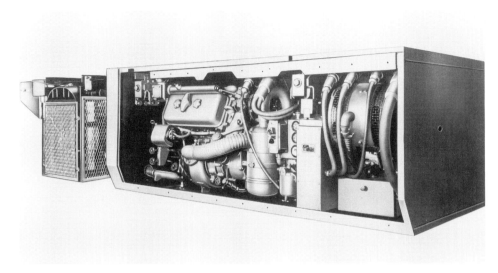

The GM diesel generator set used on the *Train Of Tomorrow*. GM.

these independent power units, as compared with axle-driven systems, reduces the locomotive trailing unit load on a 12-car train as much as 600 horsepower, 3,500 pounds or more in battery weight, gives full and constant electric output regardless of whether the car is rolling or still, increases reliability, reduces fuel cost, and also minimizes maintenance expense."[10]

"Electric power for emergency lighting and engine starting is supplied from a 32-volt storage battery, charged from a direct current exciter belted to the main engine drive shaft."[11] "Mounted on the side of the split-alternator housing, (the exciter is a) 38-volt, 1,800 rpm direct current generator providing two to three kW as required."[12]

A 40 kW, 4-cylinder diesel generator supplies 230-volt, 60-cycle, three-phase power to

A generator set shown on its pull-out rack for ease of maintenance. Pullman Technology, Inc.

The radiator for the diesel generator. Notice the standby power receptacles for 220-volt power to the right of the radiator. UPRR.

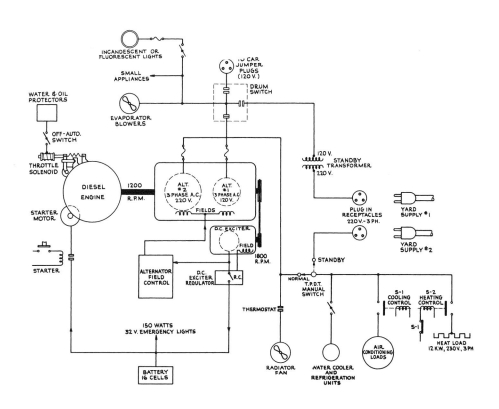

This is the block diagram for the power and air-conditioning units for three of the four cars. GM.

TECHNICAL INFORMATION

A separate 40 kW generator supplied power on the dining car, *Sky View*, for the all-electric kitchen. It, too, was mounted on a slide-out rack for ease of maintenance. Pullman Technology, Inc.

the all-electric kitchen and pantry. The unit is mounted in the car body at the rear end of the car. As with the other power units, the larger alternating unit is mounted on a sliding tray to ease repair and maintenance. A 50-gallon tank supplies fuel, and the tank is also equipped with a level gauge, liquidometer, and emergency shutoff system. The engine compartment is heated by a steam heat coil in the event the temperature drops below 50 degrees. Like the smaller units, a 32-volt battery is used to start the engine. Controls are located both inside the car and inside the engine compartment.

A 120-volt, three-phase electric trainline is provided at the end of each car, except the observation end, and a complete intertrain connector is provided for each of the cars.

Equipment Boxes

For the storage of spare parts and equipment, there are two equipment boxes on each car.

Flooring

The false floor of the lower level under the Astra Dome is low alloy steel, while the subfloor at the ends of the car is aluminum. Poplar furrings cover the subfloors. Then aluminum arched (corrugated) flooring is laid crosswise of the car and covered with Magnesite composition.

The floor of the Astra Dome compartment is made of expanded metal lathe and has the same composition flooring as the other floors in the car.

The floor in the kitchen slopes from the side to a gutter in the center. The entire floor is covered with metal pan material that is flashed up the walls and partitions to aid in draining into the center gutter. Drain holes are provided in the center gutter and are placed so as to prevent water from dripping on equipment underneath the car. The metal pan is welded to the floor, essentially forming one piece.

Like the floor in the kitchen, the floors in the pantry on the dining car and in the bar section in the observation car are covered with metal pan and designed with a similar drainage system of holes.

Floor racks covered with anti-slip Martex cover the walking spaces in the kitchen, pantry, and bar.

Floor Support in the Astra Dome

"The floor support is of high-strength low alloy steel, double sheet with spacers between. Sheets and spacers run crosswise of the car with the sheets continuous in one piece across the width of the car. Sheets and spacers are offset diagonally downward so as to form a well providing for a longitudinal passageway in the Astra Dome. Steps are provided on each side of the passageway. Suitable attachment to the car framing at the sides is provided. The top sheet is reinforced with corrugated stiffeners spot-welded into place. Spacer sections are Zee sections and I-sections. The complete assembly is arc-welded and riveted together. The floor structure is supported at the ends by means of vertical transom plates extending the full width of the car at the dashboard end of the Astra Dome. At the stairway end, the transom plates are extended from the sides of the car to vertical members located on each side of the stairway. These vertical members extend from the underframe below to the transom plates in the Astra Dome. Suitable connections are made by means of arc welding and riveting for both the transoms and uprights. On the dining car at the pantry end of the Astra Dome a vertical upright extends from the underframe providing support at this point. At the ends of the floor structures are made to conform to the sloping ceilings in the lower levels to suit conditions for each car."[13]

Framing

The general framing for all the cars is welded girder-type construction. The structural members on the underframe, sides, roof, and ends of each car are made of high-strength low-alloy steel.

Glass

All the glass in the side windows of the train, except in the vestibule side windows, in the washrooms, and the curved glass in the observation end of the observation car are double Thermopane glass. The glass in the observation end of the observation car is single unit Thermopane glass curved to fit the contour of the body. The side glass in the vestibule doors is safety glass, as is all the glass in the end doors.

The ceiling windows and the windshield glass in the Astra Dome compartments are double Thermopane with a neutral-tinted plastic laminate added in the construction. The side glass in the domes is of a double Thermopane design.

Each unit of the Thermopane design has a heat-absorbing exterior pane and a pane of safety glass on the interior. "Regular Thermopane is an insulating unit consisting of two panes of glass separated by a dehydrated air space hermetically sealed in at the factory by a metal-to-glass bond. The Thermopane glass units used on the train are made up as follows:

> The outside sheet is heat-absorbing, glare-reducing polished plate glass and is tufflexed (heat treated or tempered, making it three to five times stronger than regular plate glass) to assure extra strength.

TECHNICAL INFORMATION

The inside glass, separated from the outside sheet by a quarter-inch dehydrated insulating air space, is made up of special laminated safety glass (two panes of glass with an inner layer of plastic)."[14]

From time to time, the piece of plastic in the laminated safety glass has a slight tint to it, which is the case with the glass in the roof of the Astra Dome. This tint helps to cut glare and aid with heat reduction.

The upper part of the partitions between the three rooms in the lower level under the Astra Dome of the chair car is made from horizontally fluted glass.

At the rear of the dining car, between the main dining room and the passageway on the generator side of the car, is a partition containing polished plate glass.

Heating System

Fin type design radiant steam heating units are used throughout the train. Controlled units are arranged at floor level for each portion of the cars, with individual units for each room in the sleeping car.

Hoppers

Toilet fixtures (hoppers) made of vitreous china with vacuum breakers, flush connections, and flushing valves are located in the following areas:

- Men's toilet and dressing room in the chair car
- Women's toilet and dressing room in the chair car
- Men's and women's toilets in the observation car
- General toilet and in the annex drawing rooms D and E on the sleeping car
- Crew's toilet in the dining car

Stationary stainless steel hoppers with pedal-type flushing mechanisms are located in the upper and lower roomettes.

Folding stainless steel hoppers using air pressure water flushing activated by pedal-type flushing mechanisms are located in the compartments of the sleeping car.

Hot Bearing Detectors

The Westinghouse Air Brake Company provided one hot bearing detector for each car at no cost. Amber lights, located in the car, are visible at all times, indicating normal operation. Red indicator lights located in equipment lockers indicate trouble in the journal boxes and are visible only when the locker door is opened.

Insulation

The floors, roof, and ends have 3 inches of fiberglass insulation, while insulation in the sides measures 2½ inches.

Interior Finish

"Unless otherwise specified, aluminum is used for all interior finishes so as to reduce the weight of the cars as much as possible."[15] Most of the partitions throughout the train are made of half-inch plywood covered with aluminum. Partitions in the kitchen and pantry on the dining car and in the bar area for the observation car are covered with stainless steel.

Kitchen and Pantry Equipment

The all-electric kitchen is supplied with the following equipment:

> Three ranges with manually controlled burners and automatically controlled ovens
> One broiler interlocked with one of the ranges so that only one or the other can work at a time
> One fry kettle
> One food (steam) table
> One garbage disposal unit
> One food mixer
> Plate and cup warmers

The pantry is also all-electric and has the following equipment:

> One garbage disposal
> One glass washer
> Two 2½-gallon coffee urns
> One fruit juice extractor

The service area in the Astra Dome is equipped with the following electrical appliances:

> One four-slice toaster
> One two-burner coffee warmer[16]

Lading (in pounds)

	Empty weight	Loaded weight
Locomotive	318,000[17]	318,000
Chair car	147,000	159,000
Dining car	164,000	184,000
Sleeping car	150,000	158,000
Observation car	141,000	158,000
Total train weight	920,000	977,000[18]

	Pounds
Chair car—*Star Dust*	
Passengers and crew members with luggage, each	175
Equipment (estimate)	400
Provisions	None
Water (air-conditioning) approximately—171 gallons	1,425
Water (air pressure system)—200 gallons	1,666
Fuel oil—100 gallons	750
Dining car—*Sky View*	
Passengers and crew members with luggage, each	175
Equipment (estimated)	2,000
Provisions	3,200
Water (air-conditioning) approximately—171 gallons	1,425
Water (air pressure system)—510 gallons	4,250
Fuel oil—150 gallons	1,125

TECHNICAL INFORMATION

Sleeping Car—*Dream Cloud*
Passengers and crew members with luggage, each	175
Equipment (estimated)	800
Provisions	None
Water (air-conditioning) approximately—171 gallons	1,425
Water (air pressure system)—200 gallons	1,666
Fuel oil—100 gallons	750

Observation car—*Moon Glow*
Passengers and crew members with luggage, each	175
Equipment (estimated)	800
Provisions (estimated)	700
Water (air-conditioning) approximately—171 gallons	1,425
Water (air pressure system)—200 gallons	1,666
Fuel oil—100 gallons	750

Lighting

Fluorescent fixtures provide most of the lighting on the train. Incandescent lighting is used for emergency lighting and marker lights that operate off the 32-volt storage batteries.

Loose Car Equipment

All the cars are equipped with
- One fire extinguisher
- One flashlight
- One set of wrecking tools
- One set of car keys
- Ladders and long handled brushes for cleaning the Astra Dome glass
- Loose furniture
- Spare bulbs and fuse plugs
- First aid kits
- Vacuum cleaners
- Carpet runner
- Portable loudspeakers and microphone
- Passenger loading ramps and platforms
- Cables for standby service
- Cables for battery charging
- Cables for Station Stop City Plug-in Telephone service
- Rubber cushions for the locomotive side doors and cab headers
- Rubber cushion for the observation end door
- Portable marker lamps and spare parts
- Nylon ropes and tabs for securing ropes during exhibition

The following equipment is for the chair car only:
- Porter's step box
- Section tables
- Coat hangers for the closet next to the staircase to the Astra Dome
- Car numbering signs
- Headrest covers for the coach seats, including the dome

The dining car is supplied with the following equipment:

> Coat hangers for the crew's locker
> Loose bread, pastry, and meat-cutting boards in the kitchen, pantry, and Astra Dome
> 50 stainless steel clips to hold the tablecloths in place during exhibitions

Equipment and supplies for the sleeping car include

> Car numbering signs
> Mattresses
> Pillows—two for each berth and one for the porter's section
> Coat hangers—four in rooms—two in roomettes
> Berth ladders—one for each drawing room
> Section tables
> Folding card tables and folding chairs
> Mop and handle
> Wash bucket and wringer
> Ice bucket
> Dust pan
> Broom—parlor size
> Combination berth and water-valve key
> Hexagon wrench for the water tank
> Thermometer
> Porter's step box
> Deodorizer jug
> Headrest covers for the coach seats in the dome

Supplies for the observation car include:

> Loose cutting board at the bar
> Smoking stands
> Rug—4 feet by 6 feet—color gray—used outside the train at the observation end during exhibitions[19]

Paint

The exterior is painted in blue-green DuPont Dulux enamel. Interior surfaces are also painted with the Dulux system in various colors.

Radio, Public Address, and Wire Recorder

The train is equipped with an RCA Victor sound system that includes a broadcast receiver (AM/FM radio) that has the capability of driving speakers on all four cars, a public address system that uses the same speakers as the radio, a wire recorder capable of providing continuous entertainment over the public address system, and an antenna mounted on the roof of the observation car. Booster amplifiers located in each car drive the speakers. Speakers are located throughout the lower levels of all the cars of the train, with speakers mounted in the Astra Dome of the chair car, sleeping car, and observation car. Each speaker has a three-way switch that can be switched for the radio or the wire recorder or off. The system is mounted in a cabinet at the writing desk in the observation car. The microphone for the public address system is stored in the staircase leading to the Astra Dome. Activation of the microphone automatically cuts out the sound from the

TECHNICAL INFORMATION

The electronic equipment located in the cabinet above the writing desk in the observation lounge. Equipment includes an AM-FM radio, intratrain intercom, public address system, speaker controls, amplifiers, mobile radiotelephone, and wire recorder. UPRR.

radio or wire recorder and transfers it to the paging system. An intercar connector is provided for each car for the electric trainline receptacle at the end of each car.

The system also supports portable loudspeakers. The five speakers fit into the marker light brackets on the exterior of the cars. One speaker is put at the end of each car, on one side of the car only, except at the observation end where one speaker is put on each side of the car. A portable microphone with a 15-foot extension cord is used on the outside of the observation end during exhibitions. The speakers are painted the same color as the exterior of the train.

Refrigeration

The refrigerator and freezer units have a total capacity of 111 cubic feet, or the equivalent of 15 standard home refrigerators. The ice-making equipment can make 225 pounds of ice every 24 hours. Refrigerators used in the dining car are of a Pullman-Standard design, with the mechanical parts made by Frigidaire. The refrigerator in the bar section of the observation car is a Frigidaire unit. The refrigeration equipment includes condensing units, motors, fan and pulley assemblies, belts, and evaporators.

Using four condensing units, the refrigerators located in the dining car include

> Ice freezer located in the passageway and food storage in the kitchen
> Counter refrigerator in the kitchen
> Frozen food storage in the kitchen
> Fish storage in the kitchen with three perforated drawers
> Counter refrigerator in the pantry
> Salad counter in the pantry with a drain and a cover designed to stay open during serving times
> Ice cream well in the pantry for bulk ice cream
> Service refrigerator in the Astra Dome
> Service refrigerator in the pantry
> Ice cube storage in the pantry with two baskets or trays
> Beverage refrigerator in the steward's section with corrugated shelves for horizontally stacking bottles three high[20]

The exposed exterior surfaces of all the fixtures in the kitchen and pantry are stainless steel, while the unexposed surfaces are made with galvannealed steel. The exposed surfaces on the exterior of the fixtures in the passageway, the service refrigerator in the Astra Dome, and the beverage refrigerator in the steward's section are painted steel.

Roofs

The roof, except for the Astra Dome, is of the Turtleback type with a height from rail to roof of 13 feet 6 inches. The Astra Dome is raised 24 inches above the standard roof line and is 15 feet 6 inches from rail to roof. All the Astra Dome compartments are the same size on all the cars, with glazed flat glass sealed into structural members. The structural members are so constructed that the glass units can be applied from the outside of the car.

Seating Capacity

	Astra Dome	LowerLevel	At Ends	Total
Chair Car	24	20	28	72
Dining Car	18	10	24	52
Sleeping Car	24[21]	6[22]	14[23]	44
Observation Car	24	10	34	68

Standby Power

Standby power receptacles are located on both sides of each car to provide power for both lighting and air-conditioning. There is no standby service available for the kitchen or pantry equipment.

There are two 220-volt, three-phase, 60-cycle AC standby receptacles to furnish power normally supplied by the no. 1 alternator for lighting, small appliances, and fan blowers. "The yard power is fed to the car from the receptacle through a three-phase, 220/110-volt Frigidaire transformer mounted underneath the car."[24] Likewise, there are two similar receptacles that provide power for the air-conditioning. The no. 2 alternator normally provides air-conditioning power. Also included in the standby power package are nine 150-foot power cables with plugs. There are two each for the cars and one for the locomotive.

TECHNICAL INFORMATION 199

Telephone

Similar to ship-to-shore radiotelephone service, a radiotelephone system from Illinois Bell Telephone Company used the AT&T urban radiotelephone system. The system is mounted in a cabinet at the writing desk in the observation car. One telephone handset is supplied with the system. The aerial for the system is an 18-inch vertically mounted piano wire capable of picking up signals up to 25 miles away from 30 major metropolitan areas that offer radiotelephone services. Both outgoing and incoming calls can be made. The transmitter for the service operates on 157.89 MHz, while the receiving signal is broadcast on 152.62 MHz.

The telephone has a button in the handle that must be pushed when the passenger wants to speak and released to listen to the other person. "To make a call, passenger lifts the handset and listens, to make sure the radio channel is clear, then presses the button for two or three seconds and releases it. The operator cuts in with the name of her city and her identification: 'Mobile service operator.' The passenger gives the number of the telephone he wants to reach and also the number of the train's phone. Usually the connection takes no longer than a standard call over regular lines. For long-distance calls, the rate is the regular person-to-person charge. An extra charge is made for local calls."[25]

There is also an intratrain telephone system (intercom) to make contact with any of the four cars of the train. Each phone can handle both incoming and outgoing calls. The on-board telephones have the following extension numbers:

Chair car—no. 3   Dining car—no. 4
Sleeping car—no. 5   Observation car—no. 6

Receptacles for the electric trainlines for both the telephone and intratrain telephone system are located at the end of each car and intertrain connectors are provided for each of the cars.

Station-stop telephone service is also available on the train through receptacles located on the observation car and the room side of the sleeping car. Interior receptacles are located in the end table in the observation end of the observation car and in compartment A of the sleeping car. Telephones are provided by the Illinois Bell Telephone Company. One 300-foot cable and one 250-foot cable are supplied with the system.

Trucks

The trucks are "four-wheel, single equalizer bar all coil spring type with outside swing hangers. The truck frame, bolster, and spring plank are made of alloy steel."[26] The trucks are equipped with vertical type direct-acting shock absorbers, but have no lateral shock absorbers.

The gauge is 4 feet, 8½ inches. The wheelbase is 8 feet.

The wheels are 36 inches with the journals 6 inches by 11 inches for all the cars.

The axles are made of carbon steel and are a "new AAR standard for rolling bearings for speed up to 100 miles per hour."[27] The axles are equipped with wheel slide control devices.

The use of outside swing hangers decreases the amount of side sway from 14 inches to 3 inches by spreading the weight of the car over a broader base. The swing hangers used on the train are 96 inches apart, instead of the standard 56 inches. The use of outside swing hangers also helps with " 'banking' on curves, in proportion to the speed of the train and the degree of the curve."[28]

Hyatt roller bearings and journal boxes are used throughout. The journal boxes are of a new design that limits the lateral movement of the axle by the use of snubbers between

The trucks used on the *Train Of Tomorrow* dome cars. Pullman Technology, Inc.

The trucks for the dining car, *Sky View*. Pullman Technology, Inc.

TECHNICAL INFORMATION

the thrust blocks and the outside of the journal box. The journal boxes are also equipped with hotbox detectors.

The trucks are painted with the same blue-green DuPont Dulux enamel as the car bodies.

Vacuum Cleaners

The train is equipped with two Electrolux vacuum cleaners with all the cleaning attachments. Electrical outlets are provided in each car for the vacuum cleaners.

Washstands

Corner lavatories made of vitreous china with push-button fixtures with combination supply and pop-up waste fittings are located in the following areas:

Men's and women's toilets and dressing rooms in the chair car
Men's and women's toilets on the observation car
General toilet in the sleeping car
Crew's toilet in the dining car

China dental bowls are also located in the men's and women's dressing rooms in the chair car.

Folding stainless steel washstands of Pullman-Standard design are in all sleeping accommodations.

Water System

Water is supplied through an air-pressure system. Set into the car framing underneath the passageways of each car, the three water tanks for the cold water supply system do not

The pump, refrigeration unit, and outlet for the drinking water dispenser on the chair car, *Star Dust*. UPRR.

require any special housing, but are wrapped with two inches of fiberglass insulation. Additional water tanks are mounted in the ceiling of the kitchen. The total capacity for each car is 200 gallons, with an additional 300 gallons available in the kitchen tanks. Each of the undercar tanks has a liquidometer gauge in the rear end of the tank. Dials for the liquidometer are located inside the car and are calibrated in gallons. The kitchen water tanks have glass water gauges. Steam lines made of copper tubing surround the water tanks to keep the water from freezing.

Hot water is supplied by a steam-pressurized instant hot water heater connected to the kitchen water tanks for the sinks in the kitchen, pantry, and Astra Dome, as well as for the coffee urns, dishwasher, and glass washer. An additional 10-gallon, 230-volt, single-phase electric hot water heater is used to furnish hot water when steam is not available. A separate faucet for the electric hot water heater has an outlet near the ranges.

Hot water is supplied to the chair car, sleeping car, and observation car as well as to the crew toilet on the dining car by means of a hot water jacket being placed in the steam coil in the tank compartment underneath the cars.

A Frigidaire electric water cooler is built into the staircase leading to the Astra Dome for the chair car. Water is supplied from the tanks underneath the car.

A circulating drinking water system is built into the sleeping car. Water is refrigerated by a Frigidaire unit and then circulated by a pump located in the porter's section. A cup dispenser and a waste cup receptacle are located near each drinking water outlet in the drawing rooms, compartments, and roomettes.

The dining car and observation car do not have water coolers for drinking water. Water is chilled by ice and served by the waiters.

GLOSSARY

Astra Dome: The enclosure rising above roof level of the cars, glassed-in on all sides and top, affording all-direction vision and scenic enjoyment for 24 passengers (except in the dining car, which accommodates 18).

Bolster: 1. A long, often cylindrical, cushion or pillow for a bed, sofa, or chair. 2. The upper transverse member of a truck that holds the truck's center plate and receives the car's weight. 3. The lower transverse member of a car body to which the body center plate is attached.

Center sill: Beams that run lengthwise and through the middle of the car's undercarriage, taking the buff and drag loads of the train.

Diesel, 2-cycle: Internal combustion engines in which an injected charge of fuel is fired by the heat generated by the high compression of air within the cylinder. Two-cycle diesel engines require only two strokes of the piston (one up and one down) for each power impulse as contrasted with four-cycle operation requiring four strokes for each power pulse. The result is that two-cycle engines are more compact and provide greater power for their size.

Dummy end: The end of a car without a vestibule, but usually it has a door that permits passage from one car to another.

Electrical power package: A compact, self-contained 2-cycle diesel engine and generator for each car, providing a reliable and independent source of power for air conditioning, heating, lighting, and refrigeration.

End wall: The interior side of the exterior wall that runs the width of the car.

Es Es material: A combination of synthetic resins with wearing qualities several times greater than any linoleum or similar products; suitable for floors or walls. Available in various colors.

Flutex glass: Ribbed glass, used for panels.

Formica: A plastic impregnated with cellulose fiber.

Freon: An odorless, nontoxic, and nonflammable refrigerant developed by General Motors.

Galvannealed steel: Steel coated with a zinc-iron alloy by the hot-dip process, which consists of passing cold reduced sheet steel in coil form continuously through a pot of molten zinc. As the steel leaves the pot, the amount of coating allowed to remain on the sheet is controlled by coating rolls, which are used to establish minimum coating weights. The zinc coating is what protects the steel against corrosion.

Hot boxes: Overheated journals, caused by inadequate wheel bearing lubrication or mechanical flaws and resulting in a significant increase in bearing friction causing wheel bearing temperatures to increase. Abnormally high temperature levels result in bearing failure, which can cause derailments, endangering life, destroying property, and resulting in costly delays. Hot box detectors are thermal sensors to detect hot bearing cases. Since the bearings are on the ends of the axles, the sensors are located on the outside. The electronic hot box detector was developed to detect overheated journals on moving trains prior to bearing failure.

Inside wall: The wall inside the car, running lengthwise of the car (parallel to the outside wall of the car) separating a compartment, room, cabinet, or partition from the passageway or aisle.

Journal boxes: An antifriction mechanism at the axles.

London Glaze: A fine grain finish for leather prepared for flat wall coverings.

Oak Flexwood: Thin veneers of oak on sturdy cloth backing, flexible for application on curved surfaces.

Outside swing hangers: Swing hangers cradle the coil springs that carry the weight of the car. Outside swing hangers are mounted 96 inches apart at the lower end on the outside of the truck frame, as compared with 56 inches of separation on the inside of the frame on early truck designs. Outside swing hangers help to reduce body sway or roll on curves from 28 inches on standard car trucks down to 6 inches.

Outside wall: The interior side of the exterior wall that runs the length of the car.

Partition: A half-wall that runs from the floor up to a distance one-third to one-half the height of a normal wall.

Thermopane: Windowpanes consisting of two sheets of glass separated by dehydrated air space hermetically sealed to increase the efficiency of air conditioning in summer and heating in winter. On the *Train Of Tomorrow*, the outer glass was heat-absorbing and glare-reducing, heat-treated for extra strength. The inner sheet was a special laminated safety glass (a double sheet with a middle layer of plastic).

Transverse wall: The wall inside the car, running crosswise of the car (parallel to the end wall of the car) separating compartments, rooms, cabinets, or partitions from each other or from the passageway or aisle.

Varlar material: Stainproof plastic-coated wallpaper.

V-Board: A laminated plastic with a hard nonporous surface.

Velvean Leather: Smooth-finished leather for wall coverings.

Vestibule end: The end of a car that has a vestibule area that allows passengers to board by way of a set of stairs or directly from a platform. The vestibule area also has a door to allow passage from one car to another. Some cars (not those on the *Train Of Tomorrow*) have vestibules on both ends of the car.

Wainscot: The lower three or four feet of an interior wall when finished differently from the remainder of the wall. Usually this finish is wood, but in the case of railroad passenger equipment, stainless steel is typically used on the lower half of the wall in high traffic areas.

Weltex serrated wood: Plywood panels scraped with a knife blade with an irregular-toothed edge to produce a rough surface.

NOTES

1. "If They Could See What I See..."

1. GMC, press handout, 5.
2. GMC, proceedings of the inaugural ceremonies, 7–9.
3. GMC, press handout, 6.
4. *The Essence of Progress: The Story of the Evolution of the General Motors Train Of Tomorrow*, GMC pamphlet, 1947, 3.
5. GMC, proceedings of the inaugural ceremonies, 10.
6. GMC, press handout, 9.

2. The $100,000 Model Train

1. *Sky View* was the only dome diner built by Pullman-Standard.
2. GMC, proceedings of the inaugural ceremonies, 11.
3. The Budd Company had no relationship to CB&Q's Ralph Budd.

3. Excuse the Dust

1. GMC, proceedings of the inaugural ceremonies, 14.
2. Electro-Motive Division, GMC, press release, July 1946, 1.
3. GMC, handbook for *TOT* exhibiting staff.
4. Pullman-Standard, specification book for the General Motors Corporation's *Train Of Tomorrow*, mimeograph, July 1946, 6, 22.
5. Ibid., 4.
6. E. Preston Calvert, Michigan City, Ind., interview, September 19, 1987. Calvert worked for Carl Byoir and Associates, the public relations firm for Pullman-Standard at the time the *Train Of Tomorrow* was built. He worked with Pullman-Standard as a public relations representative on their new trains, including the GM *Train Of Tomorrow* during its eastern tour of the United States. He eventually joined Pullman-Standard, retiring as the vice president of Pullman Incorporated before it closed.

4. Something for Everyone to See

1. "*Train Of Tomorrow* Schedule," *Trains*, June 1947, 4.
2. "That Train," *General Motors Acceptance Corporation's News and Views*, June 1947, 12.
3. Ibid.

4. GMC, proceedings of the inaugural ceremonies, 7–9.

5. Paul Hampson, "Dream *Train Of Tomorrow* Is Christened," *Chicago Tribune*, May 29, 1947, 5.

6. Edward A. Braken Jr., Beverly Hills, Mich., interview, July 14, 1988. Beginning with General Motors in 1937, Eddie Braken (not to be confused with the actor Eddie Bracken) enjoyed an illustrious career with the public relations department, working on many famous GM projects and "shows." After a two-year stint with the Parade of Progress, Braken moved to New York to work on the 1939–40 World's Fair. Following that, Braken went back on the road on the Previews of Progress, an updated version of the Parade of Progress. Around 1950, the next and last stop was Detroit, where Braken was put in charge of Special Projects, which meant he had public relations responsibilities at the second New York World's Fair, World's Fairs at Seattle and San Antonio, as well as at annual meetings and other GM projects.

7. Calvert interview.

8. Dick Terrell, Naples, Fla., interview, May 14, 1988. Terrell started his career with General Motors as an apprentice in 1937, working his way up through the ranks to become a vice chairman of the board before his retirement in 1979. In September 1959, he was made general manager of the Electro-Motive Division and a vice president of GM.

9. Pullman-Standard, *Train Of Tomorrow: Built by Pullman-Standard*, brochure, 1947, 2.

10. Horace Sutton, "A Note on Trains," *Saturday Review of Literature*, July 19, 1947, 25.

11. Lloyd Unsell, "New Title Goes to GM Train—Swift Pony with Glass Saddle," *Tulsa World*, March 24, 1948.

12. Among the many artifacts that relate to the *Train Of Tomorrow* at the Utah State Railroad Museum is what many believe is Fish's original score and some handwritten orchestrations. Hundreds of hours of research over a 20-year period went into looking for a copy of the score or sheet music in music libraries and repositories, including the Library of Congress. Even though recordings were made and distributed to radio stations for play, chances are the music and lyrics were never copyrighted; therefore, a registration copy was not required by the LOC. The score in the collection became available through an eBay® online auction in 2005. Charles E. Kinzer, a devoted *Train Of Tomorrow* fan, made the winning bid and generously donated the composition to the Ogden Union Station Foundation.

5. For Sale

1. Class 1 railroads are the major railroad companies.

2. GMC, *Train Of Tomorrow* disposal program, Detroit, June 1949, 4.

3. G. Metzman, New York Central System; A. E. Stoddard, Union Pacific Railroad; J. D. Farrington, Chicago, Rock Island and Pacific Railroad Company; H. C. Murphy, Chicago, Burlington & Quincy Railroad Company; R. L. Williams, Chicago & North Western Railway Company; Judge W. McCarthy, Denver & Rio Grande Western Railroad Company; F. J. Gavin, Great Northern Railway Company; P. J. Neff, Missouri Pacific Railroad Company; A. K. Atkinson, Wabash Railroad Company; D. V. Fraser, Missouri-Kansas-Texas Railroad Company; C. Hungerford, St. Louis–San Francisco Railway Company; W. N. Deramus, Kansas City Southern Railway Company; L. B. Tigrett, Gulf, Mobile & Ohio Railroad Company; W. C. Vollmer, Texas & Pacific Railway Company; J. B. Hill, Louisville and Nashville Railroad Company; W. S. Hackworth, Nashville, Chattanooga & St. Louis Railway; F. G. Gurley, Atchison, Topeka and Santa Fe Railway System; C. E. Denney, Northern Pacific Railway Company; A. T. Mercier, Southern Pacific Company; F. B. Whitman, Western Pacific Railroad Company; C. H. Buford, Chicago, Milwaukee, St. Paul and Pacific Railroad Company; W. S. Franklin, Pennsylvania Railroad; R. B. White, Baltimore and Ohio Railroad System; E. S. French, Boston & Maine Railroad; E. E. Norris, Southern Railway System; C. McD. Davis, Atlantic Coast Line Railroad; L. R. Powell Jr., Seaboard Air Line Railroad Company; W. A. Johnston, Illinois Central Railroad Company; and R. C. Vaughn, Canadian National Railways.

6. Going to Work

1. Charles F. A. Mann, "Union Pacific Again Sets Pace," publication unknown, December 1950, 36.

2. Courtland Matthews, "Astra Domes in the Northwest," *Trains*, April 1951, 43.

3. The Milwaukee Road, memorandum from R. F. Johnson, general passenger agent, Chicago, December 30, 1957.

4. Union Pacific, memorandum from C. H. Saltmarsh, Portland, Ore., September 24, 1962.

5. Union Pacific Railroad, record of rolling stock equipment: *Sky View*, 8010, Omaha, April 1950–February 1965.

6. Union Pacific Railroad, record of rolling stock equipment: *Dream Cloud*, Omaha, April 1950–February 1964.

7. Union Pacific Railroad, record of rolling stock equipment: *Star Dust*, 7010, Omaha, April 1950–November 1965.

8. Union Pacific Railroad, record of rolling stock equipment: *Moon Glow*, 9015, Omaha, April 1950–March 1965.

7. Life after Death

1. David Randall, *Streamliner Cars*, 1:546.
2. Henry Fernandez, Pocatello, Idaho, interview, December 18, 1988.
3. Daniel B. Kuhn, "My 'Discovery' of the *Moon Glow*," handwritten ms., Portland, Ore., October 29, 1988, 3.
4. Ibid., 4–5.
5. Lester Tippe, Salt Lake City, Utah, interview, April 28, 1988.
6. Letter of donation and receipt, Henry's Scrap Metals, Pocatello, Idaho, December 27, 1984.
7. Murl Rawlins Jr., Heber City, Utah, interview, December 18, 1988. A rip track is used for storing railroad equipment or as a place to put freight cars for cargo to be loaded and unloaded.
8. Kuhn, "My 'Discovery' of the *Moon Glow*," 10.

8. Locomotive 765

1. GMC, "*Train Of Tomorrow* details," Detroit, May 1947, 2.

Appendix B

1. Pullman-Standard, specification book, 50.
2. Ibid., 56.
3. Ibid., 6–8.
4. Pullman-Standard, specification book, 10.
5. The glassware pattern is still available from Libbey Glass and can be ordered through commercial distributors.
6. Rudy Morgenfruh to author, December 28, 1988. Morgenfruh and Arthur L. Dominy are the authors of *Silver at Your Service*.
7. Pullman-Standard, specification book, 11.
8. GMC, press handout, 43.
9. Pullman-Standard Car Manufacturing Company, *Train Of Tomorrow: Built by Pullman-Standard*, brochure (Chicago, 1947), 6.
10. Ibid., 6.
11. Pullman-Standard, specification book, 52.
12. GMC, press handout, 43.
13. Pullman-Standard, specification book, 13.
14. GMC, press handout, 52–53.
15. Pullman-Standard, specification book, 22.
16. Ibid., 60–62.
17. Loaded weight. The empty weight is unknown.
18. GMC, handbook for *TOT* exhibiting staff, 11, 12–13.
19. Pullman-Standard, specification book, 96–98.
20. Ibid., 26–28.
21. Seating capacity.
22. Sleeping capacity.
23. Sleeping capacity
24. Pullman-Standard, specification book, 43.
25. GMC, handbook for *TOT* exhibiting staff, 17.
26. Pullman-Standard, specification book, 106.
27. Ibid., 107.
28. GMC, handbook for *TOT* exhibiting staff, 20.

BIBLIOGRAPHY

Magazines

"Again the Fair." *Newsweek*, July 4, 1949, 62.
"Astra Dome Train: Radical New Cars Give Passengers a Fine View and a Sway-Proof Ride." *Life*, June 23, 1947, 51–52, 54.
"Dream Clouds." *Time*, June 2, 1947, 84–85.
"Forecast for Our Future." *Pullman-Standard Car Manufacturing Company, Inc.'s Car Builder*, August 1947, 4–11.
Gardner, W. A. "A Reputation for Reliability." *Trains*, January 1979, 48–51.
"General Motors: *Train Of Tomorrow*." *Newsweek*, June 9, 1947, 72.
"Glamour on Wheels." *Newsweek*, August 2, 1948, 56.
"Glass-Topped *Train Of Tomorrow*." *Popular Mechanics*, August 1947, 111.
"GM *Train Of Tomorrow* Forecasts New Rail Era: Five General Motors Divisions Sponsor 'Shock' Promotion to Broaden Market for Equipment." *Industrial Marketing*, May 1947, 38–40.
Hawthorne, J. W. "The Indestructible Locomotive." *Trains*, January 1979, 44–47.
Ingles, J. David. "An Instant History of EMD." *Trains*, September 1972, 56–59.
Jackson, Donald Dale. "Cabooses May Be Rolling toward the End of the Lines." *Smithsonian*, February 1986, 100–111.
Mann, Charles F. A. "Union Pacific Again Sets Pace." Publication unknown. December 1950, 36–37.
Matthews, Courtland. "Astra Domes in the Northwest." *Trains*, April 1951, 41–43.
"Mechanical Innovations in the *Train Of Tomorrow*." *Modern Railroads*, June 1947, 18.
Morgan, David P. "The Essence of the E7." *Trains*, January 1979, 30–41.
Sutton, Horace. "A Note on Trains." *Saturday Review of Literature*, July 19, 1947, 25–26.
"That Train." *General Motors Acceptance Corporation's News and Views*, June 1947, 8–12, 34–35, 40.
"Tomorrow's Train Arrives Today." *Business Week*, May 31, 1947, 22.
"Train Of Tomorrow." *General Motors Acceptance Corporation's News and Views*, April 1947, 25–28.
"*Train Of Tomorrow* Astra Domes." *New Yorker*, October 11, 1947, 25–26.
"*Train Of Tomorrow* Cars Enter Besieged Northwest." *Trains*, August 1950, 10.
"The *Train Of Tomorrow* on Our Rails." *Atlantic Coast Line News*, July 1947, 3.
"'Train Of Tomorrow' Schedule." *Trains*, June 1947, 4.
"Train Promotion." *Business Week*, October 11, 1947, 32, 34.
"UP Buys General Motors *Train Of Tomorrow*." *Trains*, July 1950, 7.
"Wheels a' Rolling." *Pullman-Standard Car Manufacturing Company, Inc.'s Car Builder*, September 1948, 16–19.
"Who Had First Dome?" *Trains*, February 1954, 10.
Zimmerman, Karl. "40 Years of Domes: Part I." *Passenger Train Journal*, December 1985, 13–24.
———. "40 Years of Domes: Part II." *Passenger Train Journal*, January 1986, 13–28.

Newspaper Articles and Editorials

"Burlington's Start: A Borrowed Engine, 12 Miles of Track." *Chicago Tribune*, June 24, 1949.
Cessna, Ralph W. "Railway Riders Get Full View in Astra Dome: Sleek New Train Starts U.S. Tour." *Christian Science Monitor*, June 2, 1947.
"Christen 'Train Of Tomorrow' after Tryout." *Chicago Tribune*, May 27, 1947.
"City's Big Rail Fair Opens Gates Today." *Chicago Herald American*, May 27, 1947.
Fitzpatrick, Rita. "History Will Live Again in Rail Pageant." *Chicago Tribune*, June 24, 1949.
"GM's *Train Of Tomorrow* Here for Detroit to See." *Detroit News*, June 4, 1947.
"GM 'Train Of Tomorrow' Open for Inspection Here Today, Thursday." *Warren (Ohio) Tribune Chronicle*, October 20, 1948.
Hampson, Paul. "Dream *Train Of Tomorrow* Is Christened." *Chicago Tribune*, May 29, 1947.
Howe, Ward Allen. "Rail Notes: GM Train, Equipment with Many Innovations Will Be Exhibited Today and Tomorrow." *New York Times*, October 5, 1947.
———. "'Train Of Tomorrow': Trail Tour of Four-Level Cars Points Way to Safer, Scenic Rail Travel." *New York Times*, June 8, 1947.
"New Train Noiseless, Swayless, Three or Four Levels in Each Car." *New York Times*, May 29, 1947.
Orr, Richard. "Big Pageant Will Depict Rail History." *Chicago Tribune*, July 20, 1948.
Peck, George. "Railroads Are Here to Stay." *Paw Paw (Ill.) Times*, July 31, 1947.
Stauffer, Fred B. "GM Sells to Western Railroad Its Special 'Train Of Tomorrow'." *New York Herald-Tribune*, May 5, 1960.
"Strip Diesel Engine to Bare Secrets." *Chicago Tribune*, July 20, 1949.
"A Study in Government: American Enterprise and Governmental Dictation." Editorial, *Detroit Free Press*, June 6, 1947.
"Super Deluxe Travel: Detroit to See *Train Of Tomorrow*." *Detroit Free Press*, June 4, 1947.
"Thirty-eight Lines Join in Presenting '49 Rail Fair." *Chicago Tribune*, June 24, 1949.
"Train Of Tomorrow." Editorial, *New York Times*, June 8, 1947.
"*Train Of Tomorrow* Displayed as a Model." *New York Herald-Tribune*, February 5, 1947.
"*Train Of Tomorrow* Sold." *New York Herald-Tribune*, May 6, 1950.
Unsell, Lloyd. "New Title Goes to GM Train—Swift Pony with Glass Saddle." *Tulsa World*, March 24, 1948.

Books

Dominy, Arthur L., and Rudolph A. Morgenfruh. *Silver at Your Service*. Del Mar, Calif.: D&M, 1987.
Dorin, Patrick C. *The Domeliners: A Pictorial History of the Penthouse Trains*. Seattle: Superior, 1973.
Kratville, William. *Passenger Car Catalog*. Omaha: Kratville, 1968.
Luckin, Richard P. *Teapot Treasury*. Denver: RK, 1987.
Randall, David. *Streamliner Cars*. 3 vols. Godfrey, Ill.: RPC, 1981.

Interviews

Braken, Edward A., Jr. Beverly Hills, Mich. July 14, 1988.
Calvert, E. Preston. AD/CAL Publishing, Michigan City, Ind. September 19, 1987.
Calvert, Nancy. Electro-Motive Division, General Motors Corporation, LaGrange, Ill. November 19, 1988.
Fernandez, Henry. Pocatello, Idaho. December 18, 1988.
Rawlins, Murl, Jr. Heber City, Utah. December 18, 1988.
Rawlins, Murl, Sr. Beautiful, Utah. December 18, 1988.
Riordan, Art. Mission Hills, Kans. February 25, 1992.
Seidel, David. Columbus, Neb. December 18, 1988.
Terrell, Richard. Naples, Fla. May 14, 1988.
Tippe, Lester. Salt Lake City, Utah. April 28, 1988.
Wood, Dale. Pocatello, Idaho. December 18, 1988.

Brochures

At Ease! A Story of Travel on the General Motors Train Of Tomorrow. Detroit: GMC, 1947.
The Essence of Progress: The Story of the Evolution of the General Motors Train Of Tomorrow. Detroit: GMC, 1947.

General Motors Train Of Tomorrow. Detroit: GMC, 1947.
General Motors Train Of Tomorrow: A Brilliant Example of Industrial Teamwork. Detroit: GMC, 1947.
General Motors Train Of Tomorrow: . . . On the Tracks of Today. Detroit: GMC, 1947.
Libbey Foodservice Catalog. Toledo: Libbey Glass, June 1988.
Train Of Tomorrow: Built by Pullman-Standard. Chicago: Pullman-Standard Car Manufacturing Company, 1947.
Why We Built the Train Of Tomorrow: From a Talk by Paul Garrett, Vice President General Motors Corporation. Detroit: GMC, 1947.

Corporate Records

General Motors Corporation

Daily log for *Train Of Tomorrow* tour, May 27, 1947, through October 30, 1949.
Diagrams, floor plans, drawings, and specifications for Engine 765 of the *Train Of Tomorrow.* LaGrange, Ill., July 1946.
Handbook for *TOT* exhibiting staff. Mimeograph. Detroit, May 1947.
Press handout. Mimeograph. Detroit, 1947.
Proceedings of the inaugural ceremonies for the *Train Of Tomorrow.* Chicago, May 28, 1947.
Proceedings of the luncheon meeting for the *Train Of Tomorrow.* Philadelphia, October 9, 1947.
Proceedings of the preview luncheon for the *Train Of Tomorrow.* New York, October 1, 1947.
Train Of Tomorrow advance man's manual. Detroit, May 1947.
Train Of Tomorrow—details, typewritten ms. Detroit, May 1947.
Train Of Tomorrow disposal program. Mimeograph. Detroit, June 1949.
Train Of Tomorrow eastern itinerary. Detroit, May 1947.
Train Of Tomorrow motion picture script. Detroit, August 29, 1947.
Train Of Tomorrow western itinerary. Detroit, November 1947.

Pullman-Standard Car Manufacturing Company

Car diagrams, floor plans, and drawings for GMC's *Train Of Tomorrow.* Chicago, July 1946.
Specification book for the GMC's *Train Of Tomorrow.* Mimeograph. Chicago, July 1946.

Union Pacific Railroad

Record of rolling stock equipment. Omaha, 1950–65.

Miscellaneous

Kuhn, Daniel B. "My 'Discovery' of the *Moon Glow.*" Handwritten ms. Portland, Ore., October 29, 1988.

BOOKS IN THE RAILROADS PAST AND PRESENT SERIES

Landmarks on the Iron Railroad: Two Centuries of North American Railroad Engineering by William D. Middleton

South Shore: The Last Interurban (revised second edition) by William D. Middleton

"Yet there isn't a train I wouldn't take": Railroad Journeys by William D. Middleton

The Pennsylvania Railroad in Indiana by William J. Watt

In the Traces: Railroad Paintings of Ted Rose by Ted Rose

A Sampling of Penn Central: Southern Region on Display by Jerry Taylor

The Lake Shore Electric Railway by Herbert H. Harwood, Jr. and Robert S. Korach

The Pennsylvania Railroad at Bay: William Riley McKeen and the Terre Haute and Indianapolis Railroad by Richard T. Wallis

The Bridge at Quebec by William D. Middleton

History of the J. G. Brill Company by Debra Brill

When the Steam Railroads Electrified by William D. Middleton

Uncle Sam's Locomotives: The USRA and the Nation's Railroads by Eugene L. Huddleston

Metropolitan Railways: Rapid Transit in America by William D. Middleton

Limiteds, Locals, and Expresses in Indiana, 1838–1971 by Craig Sanders

Perfecting the American Steam Locomotive by J. Parker Lamb

From Small Town to Downtown: A History of the Jewett Car Company, 1893–1919 by Lawrence A. Brough and James H. Graebner

Steel Trails of Hawkeyeland: Iowa's Railroad Experience by Don L. Hofsommer

Still Standing: A Century of Urban Train Station Design by Christopher Brown

The Indiana Rail Road Company: America's New Regional Railroad by Christopher Rund

The Men Who Loved Trains: The Story of Men Who Battled Greed to Save an Ailing Industry by Rush Loving Jr.

Amtrak in the Heartland by Craig Sanders

Ric Morgan, a professional speaker and writer, lives in Gatlinburg, Tennessee, at the edge of the Great Smoky Mountains National Park. He is the recipient of the 2005 David P. Morgan Article Award for an article on the history and development of the dome car in *Railroad History*.

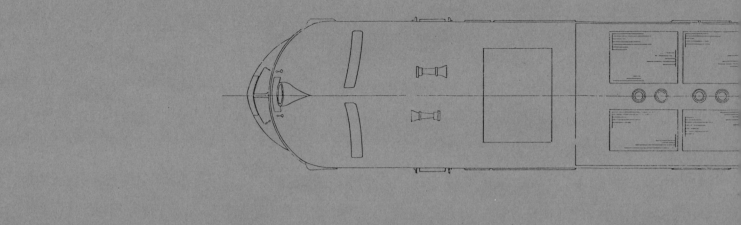
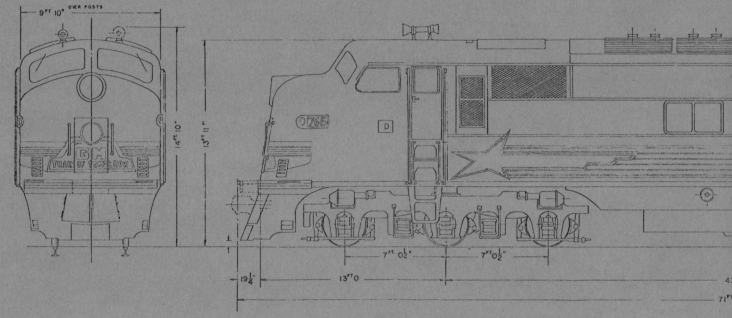
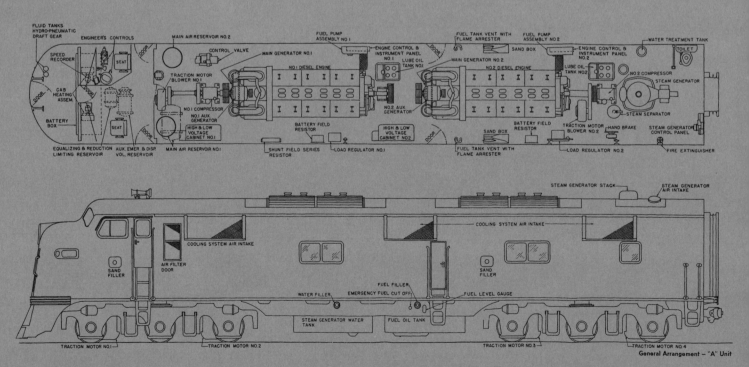

General Arrangement – "A" Unit